*Careers
for
Chemists*

Careers for Chemists

A World Outside the Lab

Fred Owens
Roger Uhler
Corinne A. Marasco

American Chemical Society
Washington, DC

Library of Congress Cataloging-in-Publication Data

Owens, Fred, 1928–
 Careers for chemists: a world outside the lab / Fred Owens, Roger Uhler, Corinne Marasco.
 p. cm.
 ISBN 0–8412–3479–5 (pbk.: alk. paper)
 1. Chemistry—Vocational guidance. I. Uhler, Roger, 1933– . II. Marasco, Corinne, 1962– . III. Title.
QD39.5.O95 1997
540'.23—dc20   96–44772
                                                                                                       CIP

The paper used in this publication meets the minimum requirements of American National Standard for Information Sciences—Permanence of Paper for Printed Library Materials, ANSI Z39.48-1984.

Copyright © 1997

American Chemical Society

All rights reserved. No part of this book may be reproduced in any form, except for brief reviews, without written permission of the publisher.

The citation of trade names and/or names of manufacturers in this publication is not to be construed as an endorsement or as approval by ACS of the commercial products or services referenced herein; nor should the mere reference herein to any drawing, specification, chemical process, or other data be regarded as a license or as a conveyance of any right or permission to the holder, reader, or any other person or corporation, to manufacture, reproduce, use, or sell any patented invention or copyrighted work that may in any way be related thereto. Registered names, trademarks, etc., used in this publication, even without specific indication thereof, are not to be considered unprotected by law.

PRINTED IN THE UNITED STATES OF AMERICA

About the Authors

Fred Owens obtained his PhD degree from the University of Illinois, working under Nelson Leonard on the chemistry of large-ring compounds.

In 1957 Fred went to work for Rohm and Haas Company in polymer research. He rose to the position of section manager in plastics research where he dealt with synthesis and applications of new polymeric systems; he obtained nine patents and authored or coauthored 10 other publications.

In 1976 Fred transitioned into an alternative career when he became manager of Research Information Services. During his tenure as manager of Information Services Fred was administrator of the on-site College Classroom program with LaSalle University and Chestnut Hill College. He was also administrator of the Teacher Training Program in which scientists took on-site courses required for Pennsylvania certification to teach high school science, a program run at Rohm and Haas by Chestnut Hill College. He also was administrator of the Delaware Valley Science program in which 400 to 600 educators learned about Rohm and Haas science and received materials which aided their teaching of science. Fred originated and administered the PROJECT LABS program in which kindergarten through twelfth grade teachers are brought into the Research Division to work with scientists to develop hands-on lessons which are used in the classroom. He also regularly gave talks on careers in chemistry to middle and high school students, college students, and at student nights at ACS local sections.

Fred was a national ACS councilor for 21 years and served on a number of Council committees. He served on the ACS Joint Subcommittee on Employment Services for 17 years and was chairman twice. He currently is a member of the Employment Services Advisory Board and is a career consultant regularly participating at national and regional résumé reviews and mock interviews.

Roger Uhler is a consultant in technology and patent management, including marketing, acquisition, and licensing under his business name Parallel Connections.

Roger retired from the DuPont Company in 1994 after 34 years in a variety of career positions: R&D, manufacturing, technical service, marketing, and in technical and marketing management. During the last portion of his DuPont career, Roger was responsible for the intellectual property (IP) management for several business units in DuPont and led a group of 12 IP specialists and support personnel.

Roger received his BS in chemistry from Pennsylvania State University and PhD in chemistry from Ohio State University. He is a Registered Patent Agent, a member of the Licensing Executives Society, and a volunteer Career Consultant with the American Chemical Society.

Corinne A. Marasco is Manager, Professional Services in the Department of Career Services of the American Chemical Society. Her professional activities cover a diverse range of topics: women in science, disability issues, corporate downsizings, and workplace issues such as pensions, fringe benefits, and harassment and discrimination.

During her tenure at ACS, Corinne has accumulated considerable experience in career-related issues. She has lectured on such topics as job security and conducting an effective job search at ACS national, regional, and local meetings. Corinne is also the editor of *WORKFORCE REPORTS*, a series providing analysis and reporting of changes and trends in chemistry's professional workforce. Her work experience has been solely in the nonprofit arena. Corinne has also worked for the American Sociological Association and the American Association of Retired Persons. Corinne received her BA degree in Sociology from the Catholic University of America and her MA in Sociology, with a focus on Work and Occupations, from the University of Maryland.

Contents

1	**Career Alternatives**	1
2	**Your Career Toolbox**	7
3	**Careers in Business**	21
	Manufacturing Management, Gary Keeney	22
	Production Management, Jamie Nickerson	23
	Business Development, Kurt Mussina	24
	Purchasing, Debra Anne Jones Stone	26
	Strategic Planning, Carol C. Segal	27
4	**Chemical Information**	29
	Patent Searcher, Stuart M. Kaback	30
	Document Analyst, Tony G. Hage	32
	Information Center Manager, Rose Ann Peters	33
	Textbook Editor, Sandra G. Kiselica	34
	Scientific Translator, Cathy Flick	35
	Business/Competitive Intelligence Analyst, Myra Soroczak	37
5	**Computer-Related Careers**	39
	Computer Modeling, M. Katharine Holloway	40
	Process Automation and Control, Joseph L. Maglaty	41
	Software Development, Neil Ostlund	43
	Software Evaluation and Documentation, Lisa M. Balbes	44
6	**Conservation of Art and Historic Works**	47
	Conservation Chemist, Beth A. Price	49
	Conservator, Whitney S. Baker	51

7	**Consulting**	**55**
	Independent Consulting, Geoffrey Dolbear	58
	Independent Consulting, Don Berets	59
	Consultant in a Large Consulting Firm, Margie Graves	61
	Consultant in a Large Consulting Firm, Allen Merriman	62
8	**Education**	**65**
	Secondary Science Education Administration, James Stockton	65
	Vice-Provost and Department Chair, Nina M. Roscher	68
	Provost, J. Ivan Legg	70
	Environmental Health and Safety, Linda W. Brown	73
9	**Entrepreneurs**	**77**
	Entrepreneur, Lyle Phifer	77
	Entrepreneur, Sallie E. Fisher	80
	Entrepreneur, Debra K. Berg	82
	Entrepreneur, Newman H. Giragosian	85
10	**Finance**	**89**
	Camp Services Supervisor, Julie A. Wurden	90
	Metals Commodity Specialist, V. Anthony Cammarota, Jr.	92
11	**Government Work**	**95**
	Law Enforcement: Forensic Laboratory, Daniel W. Vomhof	95
	Law Enforcement: State Crime Laboratory, Anna L. F. Ezell	98
	Public Safety, David R. Parker	99
	Environmental Protection, Alan M. Ehrlich	100
	County Economic Development, Heinz Stucki	102
	Armed Forces, David C. Stark	103
	National Aeronautics and Space Administration, Jack Kaye	105
	National Science Foundation, James J. Zwolenik	107
	State Department, Francis X. Cunningham	109
12	**Government Relations**	**113**
	American Chemical Society, David L. Schutt	114
	White House Council on Environmental Quality, Kathleen A. McGinty	116
13	**Human Resources**	**119**
	Human Resources Generalist, James E. Brennan	119
	Technical Recruiting, Karen Nordquist	121
	University and College Relations, C. Gordon McCarty	122

14 Law — 125
US Patent and Trademark Office, Mary C. Lee — 127
Patent Agent, Curtis P. Ribando — 129
Patent Attorney, David H. Jaffer — 131
Litigation Attorney, James C. Carver — 133

15 Medicine — 137
Physician, Donald A. Roth — 138
Nursing, Dorothy A. Bell — 139

16 Professional and Trade Associations — 143
American Chemical Society, Edward W. Kordoski — 143
Chemical Manufacturers Association, Carolyn Leep — 145
Cosmetic, Toiletry, and Fragrance Association, Joyce F. Graf — 146

17 Quality Assurance and Control — 149
Quality Control, Ron R. Richardson — 149
Process Improvement, Timothy L. Guinn — 151
Product Quality, Integrity, and Safety, Robert A. Kingsbury — 152

18 Regulatory Work — 155
Pharmaceutical Development, Bonnie Charpentier — 155
Manufacturing, Carol Jean Bruner — 157
Academic Environmental Health and Safety Manager, John S. DeLaHunt — 159
Environmental Services, Curt Clowe — 160
Clean Air (State Government), Eric Giroir — 162
Clean Water (Corporate Business), Donald P. Evans — 163

19 Science Writing — 165
Science Journalist, Ivan Amato — 166
Technical Writer/Technical Journalist, John Borchardt — 167
Public Communications, Nancy Enright Blount — 169
Public Relations Specialist, Sara J. Risch — 170
Industrial Technical Writer/Editor, Monica C. Perri — 171
Industrial Technical Editor, Gerald S. Cassell — 173

20 Technical Sales and Marketing — 175
Technical Sales, Thomas G. Tierney — 177
Technical Sales Management, Robert DiPasquale — 178
Marketing Research Analyst, Roger Walters — 180
Product Manager, Jeff Oravitz — 181

	New Business Development Manager, W. David Carter	183
	Marketing Manager, Jim Witcher	184
21	**Technical Service**	**187**
	Technical Service, Jane J. Janas	188
	Technical Service, Douglas Meinhart	189
	Technical Service, Lisa M. Headley	191
22	**Technology Transfer**	**193**
	Technology Transfer in the Governement, Richard M. Parry, Jr.	195
	Technology Transfer in the University Setting, Robert S. McQuate	196
	Technology Transfer in Industry and Business, S. J. Price	199
	Appendix: Resources	**201**
	Books and Articles	201
	Electronic Resources	202
	Job Listings and Directories	209
	ACS Career Services	211

Preface

There is a bumpy transition taking place in the workplace. The underlying assumptions about the relationship between company and employee have changed rapidly in recent years. One fundamental shift is that companies are no longer assuming the responsibility of career planning for their employees. This responsibility now rests squarely on the employees, who must now take on a more proactive role regarding their careers:

> [I]n the future chemists—like everyone else—are going to have to show more flexibility to survive and prosper. The indications are that chemists are entering into an era of more frequent job changes, fewer one-employer careers, more less-than-full-time working arrangements, more exploitation of chemistry training in nontraditional areas, and earlier career termination—either voluntary or involuntary.[1]

Being more proactive means people are in control of their careers and do not rely on an external source to manage their careers for them.

In response to these indicators, ACS first published *Career Transitions for Chemists*, a guide to help chemists who may be facing a job change or career transition. We intend *Careers for Chemists* to serve not as a sequel, but as a companion guide. In this book, we present over 80 profiles of people who are using their training in chemistry in a variety of careers. These profiles, taken from extensive telephone interviews, highlight the paths people have taken and the factors that contributed to their career deci-

---

[1] Michael Heylin. 1996. "Chemists' Employment Situation Continues To Worsen, Salaries Weak." *Chemical & Engineering News*, July 29, pp. 10–16.

sions, as well as a summary of the skills and characteristics needed for each of the 20 career fields covered.

The career information in Chapters 1 and 2 was taken primarily from ACS career publications and information presented in ACS career workshops and seminars, which are held at national and regional meetings. This information is intended to be as comprehensive as possible, but is not exhaustive.

The chapters on career fields provide an overview of each area, followed by a list of skills and characteristics. The profiles within the chapters are separated into two parts: Career Path and Career Advice. Each career profile is unique because of the different mix of skills, knowledge, and personal factors that were brought to the situation at a given time. We hope that you, too, will discover connections between your background and possible career options as you read how other chemists used their chemistry training as a springboard to a variety of career alternatives.

Acknowledgments

The ACS Board of Directors during its August 1994 meeting voted to authorize a task force to investigate the feasibility and propose the implementation of actions that would address the employment problems of chemists. The task force subsequently made a series of recommendations funded by the Board for implementation in 1995. The research for this book was one such project. The printing of the book was made possible by a 1996 Sloan Foundation grant secured by the ACS Division of Education and the Department of Career Services.

Concepts for the book were originated by Dr. Mary L. Funke, Head of the ACS Department of Career Services. Dr. Funke provided strong support for its development and completion.

We are grateful to the professionals profiled in the book who volunteered their time to be interviewed and verify their profiles before publication. They serve as excellent role models of people who take control of their careers and forge successfully forward in the uncertain world of work.

Finally, we acknowledge the fine contributions made by our editor, Karen Gulliver, who helped bring continuity, readability, and style to the book.

Fred Owens
Roger Uhler
Corinne A. Marasco

—1—

Career Alternatives

*Two roads diverged in a wood, and I—
I took the one less traveled by,
And that has made all the difference.*

Robert Frost, "The Road Not Taken"

The Changing Nature of Work

In your career travels there are many roads, crossroads, yield signs, and even stop signs. What you may choose to do at these points in your journey can make all the difference in your professional development. At one time, companies could create career paths for their employees because companies were able to make plans ten years in advance, including their staffing needs. Today, with the rapid changes in the economy and the business world, companies are divesting themselves of their career-planning responsibilities because they do not know whether they will even be in business 10 years from now, much less what business they will be in. Any training that companies do is more tactical than strategic since the focus is now on short-term needs rather than long-term objectives. This has shifted the responsibility of career planning to employees who must now take on a more proactive role regarding their careers.

Being proactive in your career means three things. First, you must take responsibility for yourself. Each individual has the responsibility for making informed decisions and implementing them. No one else can do that for you. Second, the recruiting process is changing. Fewer employers are looking for workers: They do not have to since reverse recruiting is taking place—people are looking for employers. Simply updating or changing a résumé and blanketing the country with it may only increase the likelihood of a prolonged job search, especially if your goal is a career change. Third, the route to success is no longer a straight line or a fast track. Lateral experiences can broaden perspective and skill base and make it easier to transfer skills to new markets. Skills used in one job can be valuable in other jobs. The bottom line is, *failing to recognize the necessity of being proactive can limit your options and your success.*

The Workplace Today

The underlying assumptions about the nature of work and the relationship between company and employee have changed rapidly in recent years. Employees no longer spend 25 to 30 years with one company. Employment is now provisional; the company needs the person and the person needs the job, but the arrangement is largely temporary. At one time, employees were rewarded for their performance with promotions. In today's flattened organization, the reward is more likely to be a simple acknowledgment of the employee's contribution. Management is no longer paternalistic, shepherding employees along a carefully planned career path. Instead, management empowers employees by supporting their efforts to create their own career paths under their own initiative. There are several tangible ways management supports these efforts: by giving employees permission to work on their careers; by encouraging cross-functional teams to find out what other departments are doing; and by encouraging lateral moves in place of traditional advancement. Finally, loyalty used to mean an employee would remain with one organization for the entire length of a career in return for wages, job security, and pensions. Today loyalty simply means an employee does the best job possible, with no guarantees for the future on either side.

The following employment trends were identified by the Committee on Science, Engineering, and Public Policy (COSEPUP) in a recent report on graduate education:[1]

- Among recent PhDs, there is a steady trend away from positions in education and basic research and toward applied research and development and more diverse, even nonresearch, employment.

---

[1]Committee on Science, Engineering, and Public Policy. 1995. *Reshaping the Graduate Education of Scientists and Engineers.* Washington, DC: National Academy Press.

- PhDs are increasingly finding employment outside universities and more and more are in types of positions that they had not expected to occupy.
- Over the long term, demand for graduate scientists and engineers in business and industry is increasing; more employment options are available to graduate scientists and engineers who have multiple disciplines, minor degrees, personal communication skills, and entrepreneurial initiative.

The message seems clear: chemists, indeed all scientists and engineers, need to expand their thinking outside the traditional paths of academia or industrial research in making their career decisions.

The notion of following a career outside of chemistry is certainly not a new one. US Attorney General Janet Reno entered law school after earning her BS in chemistry from Cornell University. Former British Prime Minister Margaret Thatcher found the call of politics more compelling than putting her chemistry degree to use. And who knows what drew the composer Borodin away from his chemistry studies? Despite the reduced demand for traditional chemists, a chemistry background can still be useful in the alternative career areas, particularly those skills derived from scientific training that have broader application: problem solving skills, critical thinking skills, analytical skills, and information assembly skills.[2] The ability to locate employment outside the traditional bounds of chemistry means emphasizing these complementary skills, not just technical competence.

Traditional Versus Nontraditional Work

Traditional work is typically defined as work closely tied to one's training in chemistry, such as basic R&D, manufacturing, or working at a research university. Nontraditional work can refer to work that may be related to chemistry (or science in general) but occurs in an environment other than the traditional research lab, such as patent law, forensic chemistry, or high school teaching. Alternative work occurs outside the mainstream and can represent a major departure from what people set out to do when they study chemistry. Some examples of alternative careers include scientific

---

[2]*Current Trends in Chemical Technology, Business, and Employment* is a report from the American Chemical Society based on a survey of over 100 senior-level personnel in industry, academe, and government. The report includes information on what skills employers are looking for in addition to technical competence.

translators, journalists, public policy officials, Wall Street analysts, and "hybrid" jobs such as art and book conservation where traditional work is done in a nontraditional setting.

Picture a series of concentric circles: Traditional work is at the core, nontraditional work is in the middle, and alternative work is the outside ring. The boundaries between traditional, nontraditional, and alternative work are vague and will become even more blurred as chemistry becomes even more interdisciplinary over the next decade.

Maintaining Marketability

The changing nature of work and the shift toward more diverse employment affects career management in the following ways:

- **Focus on employability**: Current employment statistics indicate that a typical employee will hold about 10 jobs with approximately 5 different employers over a lifetime. Job security does not exist anymore; in its place is the concept of *employability* or *employment security*. There is an adage which says, "The broader the base, the higher the tower." By all means, develop an expertise but do not specialize to the extent that you become pigeonholed and damage your future employability. Continue to work on your communication skills, interpersonal skills, and teamwork skills.

- **Manage your own career**: Because of the changing workplace, be prepared to respond to opportunities as well as obstacles in your career. Part of that preparation is a readiness to pursue new career options. As organizations become flatter, companies are no longer filling slots with titles but with people who have skills that are of use to the company. This means you have to begin marketing yourself as an independent contractor whose skills are for sale. When you begin to approach your job search in terms of what a company needs, what skills and abilities you have to meet those needs, and package yourself accordingly, your odds of landing an interview are likely to increase. If you can present yourself well and articulate with confidence what you have to offer, you could find yourself doing something that you never considered and enjoying it.

- **Dedicate yourself to continuous learning**: This means that your skills should be continually enhanced and updated so when a restructuring or layoff does occur, you can find a new job, be it inside the company, with another company, or in another career. Companies are not taking on the responsibility to provide training programs to keep their employees employable forever. Rather, you are now solely responsible for getting the training that will make you valuable in the marketplace and to commit yourself to a lifetime of learning.

- **Keep the inventory of your skills up-to-date:** Always be aware of your skills, strengths, and accomplishments. In the next chapter, we review the contents of your "career toolbox," including conducting a personal assessment and writing a résumé. Update your résumé yearly. Pick one day during the year, such as Labor Day or your anniversary date at work, and make this a regular activity. Keep a chronicle of your accomplishments and skills.

- **Be flexible:** When business needs change, be ready to respond quickly and adapt to those needs. Keep your options open. Expand the responsibilities in your current job by making a lateral move, taking a short-term assignment, participating in task forces, or anything else that gives you a chance to examine opportunities elsewhere.

In the end, "employability" means people are in control of their careers and do not depend on an external source to manage their careers for them. Employability also means paying attention to the job market inside and outside the company while still employed; even people with superior skills are not going to find a job if the market is not there. Adopting the outlook of an independent contractor will lead to finding more interesting, fulfilling work. The key is not to wait until you are forced into the job market to think about where your next job will be. Take stock of your skills, add to them every year, and always be ready to answer the question, "What would you do if you lost your job tomorrow?"

Changes in Latitude, Changes in Attitude

Do not fall into the trap of believing that switching to a new field will solve your current job problems, or will give you the career satisfaction that seems to elude you. As career consultant Douglas B. Richardson observed, "Most of us express desired career satisfactions in terms of what we don't want, not what we do want."[3] Avoid giving in to a second-career fantasy or magical thinking. You may find yourself in the fortunate position of having to choose between different career options, but you will feel secure in your decision only after thorough research into those options, asking yourself which skills and abilities you want to use, determining what motivates and interests you, examining your values, considering the pluses and minuses of each decision, and then acting on your information.

Using This Book

In 1994, the ACS published *Career Transitions for Chemists*, a guide to help chemists do their own career counseling, to help them discover connections between their backgrounds and work experiences, and to review those skills that are required in the job market.

In *Career Alternatives*, we present profiles of people who have used their training in chemistry as a springboard to a variety of career options. These profiles highlight the different events that led people to their present careers and factors that contributed to their career decisions, as well as any career advice they can offer.

We examine 20 career fields, beginning with an overview of the field and summarizing the skills and characteristics needed for such a career. Profiles of chemists who have gone into that field cover the career path of the individual, as well as specific advice for anyone aspiring to a similar career.

We would like to point out, however, that this book only scratches the surface of alternative careers. We are not telling you what to do because we cannot know what *you* should do. The examples in each chapter are meant to stimulate, not limit, your thinking about career opportunities.

It is important to recognize that the profiles you will read are not stories of people who woke up one morning and decided to do something different with their lives. Changing careers always involves a certain amount of risk because no one (including yourself) can ever tell which career option is "the right one" for you. Every choice we make in life influences, constrains, or eliminates other choices, now or later. Remember, there is no such thing as a sure thing.

---

[3]Douglas B. Richardson. 1995. "Don't Leap Before You Look." *National Business Employment Weekly,* October 29–November 4, pp. 8–10.

— 2 —
Your Career Toolbox

Conducting an effective job search or making a successful career transition can be challenging and rewarding if approached the right way. Just as you need the right tools to perform a scientific experiment, the same holds true when you are looking for a job. These tools—networking, résumés, and interviewing—constitute your "career toolbox."

Networking

Whether you are conducting your first job search or planning to change careers, one of the most important activities you can do is networking. Networking is not an activity that should be done as an afterthought, as a last resort, or at a time of crisis. Ideally, networking should begin at the commencement of your career (or better yet, while you're still in college) and continue throughout the length of your career.

Networking is a key activity for career success, but it is often done so poorly that it has acquired something of a bad reputation. Networking is *not* asking everyone you know for a job. There are three purposes for networking:

- ***Information***: To learn about each contact's industry and the types of jobs that exist.

- ***Ideas***: To brainstorm in order to help you develop effective job search strategies.

- ***Introductions***: To meet people who can provide more information, ideas, and additional introductions.

Research has shown that anywhere from two-thirds to three-fourths of all jobs are found in the "hidden job market"; that is, they are not currently being advertised. The only way to uncover these jobs is by networking with people who know that these jobs exist.

Everyone has a network but not everyone knows how to define or use it. A network consists of everyone you know, plus everyone *those* people know. You may associate with more than one network, each relating to an area of interest, work, or a hobby. A network includes all the people with whom you come into contact: colleagues, supervisors, clients, professors, classmates, vendors, neighbors, relatives, physicians, dentists, and lawyers, just to name a few. A network does not consist only of those people you are in contact with on a daily basis. Everyone is connected to some extent, even those people who think they have no connections at all.

There are a variety of ways to build your network:

- Join one or more professional associations, such as ACS, and attend their meetings, both national and local. Many associations also have divisions based on areas of interest; be sure to investigate those possibilities.
- Volunteer to serve on committees, task forces, or panels.
- Attend a scientific talk, or give one yourself, and introduce yourself to some of the speakers afterward to discuss their work.
- Practice small talk whenever possible.
- Participate in clubs in your areas of interest.

Networking is also taking place on the Internet. One way to network electronically is through Usenet newsgroups. Usenet newsgroups are groups dedicated to various subjects or topics that can be used to establish networking contacts, follow industry trends, current information, and specific job listings. Many of them have "jobs" in their names such as **misc.jobs.misc** (discussions about jobs and job hunting); **misc.jobs.offered** (general positions available); **misc.jobs.offered.entry** (entry-level jobs); **misc.jobs.wanted** (people looking for jobs); and **misc.jobs.resumes** (résumés can be posted here but in ASCII format only). Some newsgroups are specific to a city, state, or region, such as **tx.jobs** (jobs in Texas) or **atl.jobs** (jobs in Atlanta). In addition, there are newsgroups devoted to specific disciplines, such as **sci.chem** or **sci.physics**. Many Internet sites offer access to Usenet newsgroups but not all groups may be available.

Anecdotal evidence shows that those people who have successfully rebounded from a job loss and who have made the best career transitions had an established network to help them. By staying in touch with your network and by letting your contacts know that you are interested in job leads, you can get a jump on the hidden job market.

Networking is not the occasional phone call; it is a way of life, whether or not you are searching for a job. Some people worry about imposing on others; this concern should not stop you from networking. Be sensitive to your contacts' availability and willingness to help out, and remember that you may be in a position to help your contact in the future.

Once you have found a job, let the people in your network know and thank them for their time and help. When someone calls asking for your help in their job search, the best way to repay your network is to return the favor. Share job search techniques that have worked for you. Be attentive to the needs of others in your network and refer people who have skills another might find useful. Networking is not a one-way street; giving and receiving are part of networking. Remember to maintain your network—chances are, you will need it again.

Information Interviewing

People who want to make a career change use information interviewing to build their network and make contacts in their targeted field or career. Information interviews differ from employment interviews in a number of ways. In the employment interview, employers seek out candidates, set up the appointments, and find out what they need to know. In an information interview, *you* initiate contacts, seek out potential employers, ask questions, set up the appointments, and find out what you need to know. This is a good way to find a position, although nothing may be available at the time of the interview. Your goal is to leave a positive impression so that when something does arise, you will be remembered.

Like employment interviews, information interviews require research. You will need to ask a variety of questions about the company, the types of people who are hired there, and what they do. Do not forget to present some information about yourself: what you like to do, what you hope to do in your new career, and what skills you have that are transferable to your target field or career.

Whether you write or telephone to request an information interview, you need to focus on the following:[1]

---

[1] Martha Stoodley. 1990. *Information Interviewing: What It Is and How to Use It In Your Career.* Garrett Park, MD: Garrett Park Press.

- You are planning to change careers and want to discuss other jobs where your skills, education, and experience would be an asset.
- You have done your research and want to speak to people in the field to fill in the gaps before you start sending out résumés.
- You are new to the area and are investigating companies in the area before you launch your job search.
- You have decided on a field and are looking for feedback on your résumé and how realistic your goals are.
- You feel your talents lie in a particular field and want to find out ways to enter it.

Information interviews can be helpful and can prepare you for an employment interview. However, an information interview must be brief; know what information you are after and focus your questions accordingly. Ask open-ended questions that encourage your contacts to speak at length about themselves. Ask questions that begin with Who, What, When, Where, and How. As in any networking situation, people are usually generous with their time and advice, but remember that the purpose of the meeting is to gather information and not to secure a job interview.

If a specific job lead does not become apparent, you can turn this negative into a positive by asking for names of other people who might be able to help you. This way, you broaden your network and better target your market at the same time.

Résumés

How do you determine what you have to offer a potential employer? How do you know if you are qualified for a particular job? Your goal as a candidate is to present yourself in the best possible light, so that you stand out as the best interviewee and enhance your candidacy. You can reach this goal with some advance preparation. The place to begin is with yourself—know the skills and accomplishments you can offer an employer.

Preparing a List of Skills and Accomplishments

Why should you take time to identify your skills and accomplishments? There are many well-qualified candidates competing for each position, including positions that may be less challenging than the jobs they previously held. Clearly identifying your skills and accomplishments will help you:

- Write your résumé
- Give you confidence
- Identify the needs of the market as well as market yourself
- Answer employment interview questions, especially "What can you do for us?"

The process of identifying your skills and accomplishments is not something you can do in a few hours, but must be done over a period of time. Find a quiet place where you can think about what you have done in your personal life, work experience, or at school, in the case of new graduates. Jot down the accomplishments that come to mind, including modest accomplishments, and mention outcomes as well. Remember that results count, and be as specific as possible.

When you have listed as many accomplishments as you can generate, look at the list and think about what skills you have that allowed you to be successful. Write down whatever those skills may be: computer skills, communication skills, leadership skills, or technical skills. Assign each accomplishment to one or more of the skills categories. For example:

<u>Technical Ability in Chemistry</u>

— Initiated reevaluation of new cure agent for general powder coatings

— Designed and developed catalysts, curing agents, performance-improving additives, and agricultural and pharmaceutical intermediates

<u>Communication Abilities</u>

— Drafted written recommendations for the implementation of a patent tracking system, which were accepted by management

— Coordinated science education program with local elementary school district by teaming volunteers with science teachers

Continue to edit and modify the list until you are satisfied. When the exercise is over you will be left with a list of your skills supported by examples (your accomplishments), which you can use to write cover letters and your résumé to prepare for interviews.

A thorough assessment of skills and accomplishments is essential if your goal is to make a career transition. To undertake a successful transition, you must be able to show that you have the skills and abilities to demonstrate your competence in a new field. If you determine that your current background fits your targeted field well, you will have more confidence in making the change. If you are lacking some key skills, work on developing those skills. The American Chemical Society has published a guide, *Career Transitions for Chemists,* for those who want to undertake a job change or career transition. The principles of transition discussed in this book are useful and highlight the importance of matching one's talents with an employer's needs.

Recruiters receive hundreds, or even thousands, of résumés, but spend approximately 30 seconds reviewing each résumé. In that small amount of time, they can determine whether or not they want to have a second look at a candidate because they know the job openings and what degrees and disciplines are required. With the advent of scanning and retrieval technology, the time spent reviewing a résumé is even less. For example, identifying appropriate candidates for an opening as a polymer analytical chemist meant searching through hundreds of résumés by hand, placing an ad, or visiting colleges and universities. With a résumé database and retrieval software, the computer does all of the searching using Boolean logic (AND, OR, or NOT statements) to find candidates within the database who meet the job requirements.[2]

Résumés are summaries of information. They must be presented succinctly and in an organized fashion so that whoever is reviewing them can do so in a short amount of time without overlooking key information. A résumé lists the applicant's education, degrees, work experience, professional activities, and any honors or awards. The box on the next page summarizes what a résumé should provide a potential employer. There may be an accompanying list of patents and publications, generally considered to be an appendix to the résumé. Whether you include a list of references depends on your situation. If you are a new graduate, it is more or less expected that you will be job hunting, so including a list of references is

---

[2]Joel I. Shulman. 1995. "Making Your Resume Computer Compatible." *Today's Chemist at Work,* 4(8):43–96.

WHAT A RÉSUMÉ SHOULD PROVIDE

- *A sense of purpose:* Give the reader an indication of the type of position you seek. The reader should not reach his or her own conclusions about your job goals and your qualifications for the opportunity, because the wrong conclusions may be drawn.

- *Emphasis on achievements:* Include achievements as bulleted statements under a skills listing or, if using a chronological format, ensure that the narrative or bulleted statements reflect your achievements. Try to differentiate yourself from other applicants in a positive way (for example, showing leadership skills by professional roles).

- *Accuracy and credibility:* Never exaggerate your qualifications. Be truthful and accurate. Do not give false information or inaccurate job titles. Present your qualifications in the most impressive light; however, misrepresentation can cost you an interview, even a job.

- *Clarity and simplicity:* Remove unnecessary words to facilitate scanning of the document.

- *An attractive package—design and layout:* Leave a one-inch margin all around your résumé. The layout should be clean, with ample white space, and should not exceed two pages. Pay attention to the font you use. Fonts like Prestige Elite, Courier, Times Roman, and Geneva print clearly. Use 12-point size type so that your résumé will be easy to review.

- *Salesmanship—measurable facts that appeal:* Try to develop a match between your personal assets and those desired by the organization to which you are applying.

- *A sense of the person behind the document:* Your information about work and outside activities should convey a picture of you as a well-rounded person.

Source: *Career Transitions for Chemists,* p. 39.

permissible. If you have previous work experience, you may want to be selective in handing out your list of references. For example, you may want to tell your references more about the job before they are called upon to give a reference for you.

There are two popular résumé formats, the chronological and the functional (also called skill-based) résumés, although there are as many résumé formats as there are books written on the subject. Each format has its advantages.

The Chronological Résumé

The chronological résumé provides information on job history. It is intended to present the candidate's job titles, the levels of responsibility held in various positions, and the names of previous employers. It is most useful when a job candidate wants one or more of the following:

- To show good career progression without gaps
- To stay in the same line of work
- To enter the job market for the first time
- To highlight employment at particular firms
- To highlight the level of work activity

The chronological résumé is best used when applying to traditional organizations. The disadvantage in using this format is that it does not adequately highlight all accomplishments or transferable skills. If you have an irregular employment history, are changing career paths, or if you have changed jobs frequently yet acquired many skills in the process, you probably do not want to use a chronological résumé.

The Functional Résumé

The functional résumé highlights skills, knowledge, and accomplishments rather than job history. The functional résumé is useful for more experienced chemists when

- Changing career paths
- Seeking lower level job responsibility than held in previous employment
- Avoiding emphasis on gaps in employment or on periods of unemployment
- Diverting attention away from age
- Your experience or education does not exactly match the stated or known job requirements

The functional résumé is most useful when applying to nontraditional organizations and small companies. The disadvantage of using a functional résumé is that some employers may view a functional résumé with reservations because the skills acquired may not be in chronological order. It is best not to use this format when you are applying for traditional jobs, teaching positions, or positions in the government where career growth is important.

There are several sources to consult on writing a résumé. The ACS publications *Career Transitions for Chemists* and *Tips on Résumé Preparation* and the videotape "Developing the Right Picture: Résumé Preparation" are good resources. In addition, ACS members can request assistance from an ACS career consultant. See the section on ACS Career Services in the Resources appendix at the back of this book.

Interviewing

A résumé serves as a tool to obtain an interview, but it is the interview itself that determines whether you will receive an offer of employment. The interview gives a candidate and an interviewer the opportunity to exchange information about the position at hand. Your goal as a candidate is to present yourself in the best possible light, so that you stand out as the best interviewee and enhance your candidacy. To reach this goal, it is essential to prepare for every interview and to take it as seriously as preparing for a course examination.

As mentioned in the section above on résumés, it is vital to know your skills as demonstrated by accomplishments, and potential contributions that you can offer to an employer. It is a buyer's market; many well-qualified candidates are competing for each position, including positions that may be less challenging than the jobs they previously held. If you are seeking to make a career change, the major obstacle you will have to face is convincing interviewers that you have the skills and abilities to perform a job as well as someone who already has experience in the field you are trying to enter.

Before an interview, research the company and the industry. What you learn will aid you in demonstrating that you have the skills, abilities, and experience for which an interviewer is looking. As you study the information, you can also begin to formulate questions to ask during the interview. Few candidates realize just how important it is to become familiar with a company prior to an interview; without advance research on the company, there is no way you can satisfactorily answer the question "What do you know about our firm?"

If the company is publicly held (i.e., traded on any of the stock exchanges), you can obtain a copy of the most recent annual report from its Public Relations Department. Annual reports contain such information as the company's goals for the coming year, highlights from the preceding year, company milestones, and financial statements. You can also glean useful information about the company's officers, recent mergers or acquisitions, and about any divisions or subsidiaries.

A reference librarian can also be a valuable resource. There are a number of reference sources you can consult for information, and he or she can help you find them. Some of the more general references include *Standard & Poor's Register of Corporations, Directors, and Executives; Directory of American Research and Technology; Thomas' Register of American Manufacturers*; and *State Manufacturing Directories*. You may also find articles that have appeared in newspapers, magazines, and trade publications.

If you know someone who currently works, or formerly worked, for the company, talk to them. Faculty members may have former students now in that organization. If you know someone who has worked with the company as a vendor or consultant, they may also be able to provide you with valuable information. Try to get as much information as you can, but remember that what you hear will reflect that person's experiences, good or bad, with the company.

If you arrive at an interview having done some research about the company, you can benefit in three ways. One, you create a favorable impression by showing you made the effort to acquaint yourself with the company. Two, you have set yourself apart from the competition because so few candidates bother to research companies with whom they are interviewing. Three, you communicate a certain degree of respect for the company (and the interviewer) by not having to say, "Gee, I'm really not that familiar with what your company does...." If you do not make the effort to do some research prior to the interview, you cannot reasonably expect to receive an offer of employment. See the box on the next page for a checklist of things to do as the date of your interview approaches.

The Interview Process

During the employment interview, the interviewer will attempt to determine three things:

INTERVIEW LOGISTICS

- Practice interviewing with a friend or spouse. If possible, have someone videotape your practice interview and critique it. ACS offers taped Mock Interview Sessions at all National Meetings and some regional meetings; take advantage of the opportunity.

- Review and study your résumé, no matter how many times you have read it before. Keep your achievements fresh in your mind so you will be ready to tell the interviewer about them, and how they match the company's needs.

- Review the interview itinerary. Whom will you be seeing, and what are their titles? When will you see them? How long will you be meeting with them?

- Review the directions to the interview, how long it will take you to get there, when you plan to leave, and how you are going to get there. Jot down the time and place (including floor and suite number) so even if you are a little nervous, you will still arrive at the right place at the right time. Make a note of the interviewer's telephone number so that you can call if you get detained on the way to the interview. Bring enough change for subway fare, parking, tolls, or phone calls. Listen to the weather forecast in case you need to pack an umbrella or wear an overcoat.

- If your interview is out of town, confirm all transportation arrangements. Call the hotel to confirm your reservation and guarantee it for late arrival in case of delays. Call the interviewer to review the procedure for submitting your expenses.

- If there will be forms to fill out, ask to have them sent to you in advance of the interview. You can then take your time to fill them out completely and neatly without having to rush.

- Take along the information you have assembled on the company, extra copies of your résumé, your publication list, and your list of references, a pad of paper, and a pen or pencil to take notes during the interview.

- Bring a list of job-related questions. At some point during the interview, you will have the opportunity to ask questions of the interviewer. You might ask such questions as: "Why is the job open?" "How long has the position been open?" "How does the department fit into the company organization plan?" Do not ask a question simply for the sake of asking a question; know why you want the answer.

- Get a good night's sleep!

- Your ability to do the job; an interviewer will look for evidence that you understand how to apply what you know.

- What motivates you to work; an interviewer is looking for examples of your leadership skills and initiative.

- Whether you will fit into the organizational culture; an interviewer will want to know that you have worked well with people and will continue to do so.

Most interviewers do not cover each area equally. Since technical qualifications are more apparent, easily defined, and more easily measured, some interviewers may feel more comfortable asking a candidate about their qualifications. Unlike technical qualifications, motivation and organizational fit are more abstract concepts; it is likely that few interviewers feel sufficiently prepared to ask in-depth questions. Most interviewers may limit themselves to some basic observations about a candidate's motivation and some notion as to whether they like the candidate.

Knowing that the employer will probably focus on these areas during the interview, you can prepare by:

- Reviewing your technical knowledge, skills, and accomplishments. This includes preparing examples to show how you can use your skills in the new job.

- Communicating, both verbally and through your body language, your interest and enthusiasm for the position.

- Observing the people you meet during the interview and the environment, as well as how you might fit in or contribute.

The first few minutes of the interview are the most important because that is when your interviewer will form the first and lasting impression of you. How you look and how you conduct yourself will determine not only whether the interviewer likes you as a person, but sets the tone for the rest of the interview. There are a number of things you can do to make the best initial impression possible on the interviewer:

- Begin and end every interview with a firm handshake and look the interviewer in the eye.

- Be enthusiastic in your greeting. Smile!

- Do not call the interviewer by his or her first name, unless you are invited to do so. If your interviewer is female and you do not know her educational or marital status, address her as "Dr." or "Ms."
- Wait until the interviewer is seated or invites you to sit down.
- If the interviewer tries to break the ice with you, "fall through the ice" with him or her. Interviewers use this technique to help candidates feel at ease.

In an interview, try to spend 50% of your time listening and 50% talking. The time that you spend listening is very important. Often, interviewers will telegraph cues about the company through what they say during an interview. At some point during the interview, the interviewer will ask, "Do you have any questions?" You may be able to formulate a useful question from information gleaned during the interview.

Your communication skills are critical in an interview, especially if your goal is to enter a new field. You must be able to persuade the interviewer that you can perform the job. Do not expect interviewers to understand how your previous experience qualifies you for the job. Be ready to address the interviewer's concerns about hiring you over someone with previous experience. After all, if you do not believe your skills are transferable or cannot convince the interviewer why you feel a new career will be more satisfying, you cannot reasonably expect to receive an offer.

After the Interview

Immediately after the interview, send a thank-you letter to each person you met. Be sure to spell all names properly and use correct titles. You can ensure this by getting business cards from everyone you meet during an interview. If you had an interview out of town and the company is covering your expenses, be sure to promptly submit an expense account, including all your receipts.

If you decide that you are not really interested in the position after the interview, send a thank-you letter anyway. You may want to apply for another job with the same company in the future or be referred to another position. Also, you may end up working for a company that does business with the organization, so you want to have created a lasting good impression.

If you had a screening or site interview, thank the person for meeting with you to discuss the position. Reiterate your interest in the opening as well as your hope to work for the company.

If you had an information interview where no specific position was discussed, thank the person for their time and ask him or her to keep you in mind if a suitable position opens up. Regardless of the type of interview, do not use the thank-you letter to explain your qualifications. Keep it simple and concise. Its sole purpose is to thank them, not to sell you.

Begin Your Journey

In conclusion, you need to have your career toolbox—an established network, a polished résumé, and interviewing skills—in good order before embarking on your quest for an alternate career. Just like the traveler contemplating visiting a foreign country, however, you must also be open to the possibilities before you. It is our hope that the following profiles will prepare you for the possibilities and variety of paths that can be taken to achieve career satisfaction.

— 3 —

Careers in Business

The Committee on Science, Engineering, and Public Policy report on graduate education[1] stated that the growth sector of employment has been business and industry. In 1973, 26% of science and engineering PhDs were employed in business and industry 5–8 years after their PhD; in 1991, this proportion rose to 41%. Some chemists go into manufacturing or production management, business development, or business services such as purchasing or strategic planning. We have chosen to highlight five of these career options.

If you are interested in management and development, an MBA may be desirable. What is needed is a knowledge of business, especially finance and possibly accounting. In the service area, chemistry training and experience can help you understand what needs to be done and how to get the information needed for the job. In all areas, interpersonal skills are vital because you will be working with diverse groups of people and need to be able to handle them with tact and sensitivity. Self-motivation, being a team player, being goal-oriented, and an ability to prioritize are other key attributes.

This chapter contains a few examples of how some chemists went from the bench into business careers, including manufacturing management, production management, business development, purchasing, and strategic planning.

---

[1] Committee on Science, Engineering, and Public Policy. 1995. *Reshaping the Graduate Education of Scientists and Engineers.* Washington, DC: National Academy Press, p. 144.

> ## Skills and Characteristics
>
> - An MBA may be desirable, but not always necessary
> - A knowledge of business principles, including finance and accounting
> - Self-motivation
> - Ability to be a team player
> - Interpersonal skills
> - Goal-oriented
> - Communication skills, verbal and written
> - An ability to prioritize
> - Flexibility

Manufacturing Management

Gary Keeney

Education
BS in chemistry, Wheaton College (IL); master of management, J. L. Kellogg Graduate School of Management, Northwestern University

Career Path
Gary Keeney, vice-president of the Chemical Division, Professional Chemicals Corporation, is a general manager responsible for chemicals manufacturing and product development. Since it is a small company, his work brings him into contact with the sales and marketing, regulatory compliance, finance, legal affairs, and strategic planning functions.

Keeney obtained his BS in chemistry and became a development technician for the chemical products division of ServiceMaster Company, which franchises cleaning and business services. He became development manager, where his responsibilities included product formulation development, technical service, and training and development management. While at ServiceMaster he obtained a Master of Management degree, the equivalent of an MBA.

Through networking with sales and production people at one of ServiceMaster's suppliers, Professional Chemicals Corporation, he obtained his current position. He says that the most important training he had beyond his bachelor's and master's degrees was a Dale Carnegie course, "How to Win Friends and Influence People," he took while at ServiceMaster. He uses leadership and people skills every day. He is currently taking a nine-month course in manufacturing management in order to obtain a Materials Management Certificate to further improve his management skills.

Keeney enjoys his work and intends to remain there. If he decides to make a move, he intends to stay in manufacturing management.

Career Advice
Keeney points out that manufacturing management requires teamwork and people skills. In his capacity he works with a wide variety of people, including workers on the shop floor, technicians, salespeople, attorneys, and customers. He stresses that it is important to be goal-oriented, self-motivated, and to have the ability to prioritize.

Keeney states that it is probably not necessary to have an MBA, but says that expertise in the areas of finance and accounting is essential. He also feels that in the chemical industry, a manager must have legal and regulatory knowledge.

Keeney advises anyone contemplating a career in manufacturing management to start doing their research as early as possible, and explore taking business and management courses. It is also helpful to get some exposure to manufacturing.

Production Management

Jamie Nickerson

Education
BS in chemistry, Seattle University (WA)

Career Path
Jamie Nickerson worked part-time at a cancer research center as an undergraduate, and then full-time for a year. She obtained an entry-level position as a formulations chemist in the manufacturing department of a biotech diagnostics company. During her eight years there she held a number of

supervisory positions in manufacturing, rising to the position of production manager for a new diagnostic product. She built the department from the ground up to 14 people while designing the facility and setting policy.

Nickerson then joined DepoTech Corporation as Production Manager for a new drug in clinical trials. She works closely with the process development department and the company expects to file an application to market the drug. While the production department helps to produce other drugs in the early stages of development, it is essentially a pilot plant. DepoTech is simultaneously building a manufacturing plant which requires her input concerning the construction of the plant and planning the equipment.

Nickerson currently supervises a mix of BS biochemists and biologists, and laboratory technicians. She is working on an MBA because she feels she needs business training to better perform in her present position. Nickerson enjoys what she is doing, and particularly appreciates the freedom she has to do it her way. In five years she hopes to be at a higher level in production management.

Career Advice

Nickerson says that for a research and development job an MBA or some kind of a business training can be helpful. She feels that to do well in production requires an individual who is organized and detail-oriented. Nickerson advocates developing good interpersonal skills because it is necessary to be able to work with a wide variety of people of different backgrounds including microbiologists in the lab, chemists in the quality control lab, and warehouse workers. She finds that she must be flexible because things can change quickly in the business environment.

Nickerson says that work in production requires someone who is more logistically oriented than technically oriented—someone who can find a better, more efficient, less costly way to do things. Production is not routine and is practical; part of the satisfaction in her job comes from having something to show for her work.

Business Development
Kurt Mussina

Education
BS in chemistry, Montclair State University, Upper Montclair (NJ)

Career Path

Kurt Mussina obtained a BS degree in chemistry and worked for a small generic drug company as a laboratory chemist. He then moved to a large international pharmaceutical company as a scientist. Just as Mussina was tiring of bench chemistry, he saw an ad in his local newspaper for a Manager of Analytical Business Development at Applied Analytical Industries (AAI), an international contract pharmaceutical services company.

Mussina's job as business manager included project definition and contract development, including pricing, which depended on the complexity of the procedures and the quantity of samples. He also had to negotiate the terms of the contract in person or over the phone, and ensure that the client's material, including high performance liquid chromatography (HPLC) and gas chromatography (GC) met US Pharmacopeia and National Formulary standards.

Mussina advanced to Technical Director of Analytical Business Development, responsible for developing AAI analytical services nationwide, a position requiring client contact in person and by phone, extensive travel, making presentations, reviewing ongoing contracts, looking for new business opportunities, and discussing problems.

He has recently been promoted to a position in international business development. He is currently in Denmark and will be responsible for business development in Scandinavia, Germany, Belgium, and the Netherlands. His responsibilities will include analytical services and contract services in formulations development, biotech analyses, clinical supplies, manufacturing, and regulatory affairs consulting. In the future Mussina sees himself with an MBA and as vice-president of business development.

Career Advice

Mussina says that until this point his chemistry degree and good communication skills have been sufficient and no further formal training has been necessary. However, he does admit that an MBA would have been very helpful and he plans to get one when he returns from Europe. He states that good interpersonal skills are needed to handle clients and he notes that he has acquired on-the-job knowledge of government regulations regarding pharmaceuticals.

He advises anyone considering working in business development to get a chemistry degree and an MBA, then get experience with both a small generic and a major pharmaceutical company. He feels that with a BS in chemistry and an MBA one can get a job in the pharmaceutical industry.

Purchasing

Debra Anne Jones Stone

Education
BS in chemistry, Vanderbilt University

Career Path
Debra Stone started her professional career as a senior laboratory technician at McKee Foods Corporation where she worked for four years analyzing the ingredients of Little Debbie® Snack Cakes and ordering supplies for the lab. She gradually started ordering for the quality assurance and microbiology labs, until eventually she was ordering supplies for all of the laboratories. Management then decided to go to centralized purchasing and expanded the purchasing department. Because of a need for material safety data sheets (MSDS) for each material purchased, management also decided that the chemical purchasing agent should be a chemist. Stone got the job.

Because she is a chemist and knows what is involved in the testing of materials, Stone handles MSDS approvals and hazardous waste disposal. She also handles the contracts for outsourced testing such as that of hazardous and nonhazardous waste. She says she is able to negotiate better prices because she has run a GC-Mass Spectrometer and knows if a vendor is trying to overcharge for the service. She says it often is a surprise to a vendor that a purchasing agent is knowledgeable in chemistry. Stone is a Certified Purchasing Manager (CPM), which required that she pass four exams on purchasing. She and her manager decided that she should go for the certificate since she would learn much while studying for the exams. Stone would like to succeed her manager if and when her manager is promoted. If she ever tires of purchasing, she may move into the quality improvement area because of her proficiency in math and because she finds the use of statistics fascinating.

Career Advice
Stone says that beyond her chemistry background and her certification, a position in chemical purchasing requires people skills, negotiation skills, customer service skills, and an ability to work with the suppliers. Computer skills are also important because much of the purchasing function is automated today. She has been able to develop supervisory skills on the job.

She recommends chemical purchasing highly to anyone with a chemistry degree. The chemistry background is very helpful particularly when buying chemicals or contracting for services.

Strategic Planning

Carol C. Segal

Education
BS in chemistry, University of Chicago; PhD in physical chemistry, University of California at Berkeley

Career Path
After obtaining a PhD in physical chemistry under Nobel Prize winner Yuan T. Lee, Segal began her professional career as an assistant professor of chemistry at Swarthmore College. Four years later she returned to full-time research at The Aerospace Corporation where she worked on the chemical physics of propulsion and atmospheric chemistry in the aerophysics laboratory. She later transferred to the Survivability Directorate where she supported underground nuclear-test planning and two sounding-rocket test programs. After three years she joined the Materials Sciences Laboratory performing vulnerability analysis (analyzing what threats there are to hardware that the military has in space) of an Air Force communications satellite and research in a novel electroplating technique.

Three years later Segal was promoted to manager of the Metals and Composites Section where she developed a horizontal engineering program on spacecraft aging concerning flightworthiness of satellites and launch vehicles following ground-based storage. In 1995 she was selected as one of the Aerospace Women of the Year.

In her current position as senior scientist in the Strategic Planning Office at The Aerospace Corporation, Segal works with corporate executives and senior managers to interpret the external business environment; maintain awareness of actual and perceived corporate strengths and weaknesses; develop and communicate corporate strategies; facilitate operational planning within the corporation to implement these strategies; and provide an independent assessment of progress toward corporate goals.

Segal enjoys change, as evidenced by her career thus far and her present position, which she obtained through a posting. In strategic planning she builds on one of her strengths, writing. Her current job puts her in a corporate position with an overview of a broad range of corporate activities, and an opportunity to learn more about the entire business. She points out that this was not part of a grand career plan; the opportunity presented itself and she pursued it.

In the future Segal sees herself in the management of science, perhaps returning to managing research to help internal research support the goals of the corporation. She sees the company's future in space and wants to be a part of that.

Career Advice

As a strategic planner in the aerospace industry, Segal feels that having a technical background gives her a better understanding of what her company does. She must be able to understand the technical and business aspects of her company's contracts with the government. In dealing with projects that involve volumes of data coming in from disparate groups, she has to be able to work with, analyze, and pull together technical information and extract the essence of what is needed. Segal feels that in graduate school she developed the ways of thinking and the analytical tools she needs day-to-day to assimilate and use the data.

A valuable skill for a strategic planner beyond a PhD is an ability to make written and oral presentations that provide logical and reasonable conclusions based on data. Interpersonal skills are critical; in Segal's work she must convey to senior management the needs and concerns of middle and lower management. She gets this information by mingling with people and establishing a rapport with them. Segal has taken some courses in strategic planning. She also feels that by participating in a great books discussion group, she honed important skills needed in her current position; she read new and often unfamiliar material which she then discussed with a diverse group. Because many participants had opinions which differed from hers, she learned consensus-building.

Segal's advice to anyone contemplating work in strategic planning is that there is no clear-cut career path; having an open mind to change is important. She did not know a lot about strategic planning but recognized that she had many of the skills needed for the job, so she pursued the opportunity.

—4—
Chemical Information

Chemical information specialists handle and manage business and technical information in a variety of ways. Possible jobs include: performing searches, analyzing documents, editing textbooks, technical editing, translating from foreign languages, managing an information center, and acquiring business or competitive intelligence.

A thorough knowledge of chemistry (understanding the language, nomenclature, structures, and physical properties) and the scientific literature is vital since most chemical information jobs require reading and analyzing data of a technical nature. The information specialist must be able to understand, analyze, and summarize data so that the client or customer can grasp the information the specialist has found. An ability to organize and present information is very important.

A chemical information specialist must have a greater interest in the scientific literature than in the scientific method (experimentation). Attention to detail and scientific curiosity coupled with an analytical ability and willingness to learn are important traits. The information specialist must be a generalist yet must also maintain current knowledge in all branches of chemistry as well as a familiarity with other sciences and business.

Most information specialists emphasize the importance of being able to work with people and to communicate well orally and in writing. Computer skills, especially access to information and statistical databases, are vital.

Information specialists are needed in libraries, chemical companies, market research firms, consulting firms, and organizations which do docu-

> ### Skills and Characteristics
>
> - A master's in library science is necessary to work in an academic library and helpful in an industrial scientific library.
> - Computer use (must be able to access database vendors, such as Dialog, and use the Internet)
> - Analytical ability
> - Understanding of the language of chemistry, chemical nomenclature, chemical structures, and physical properties
> - Organizational skills
> - Good writing and editing skills
> - Interpersonal skills
> - Attention to detail
> - Flexibility

ment analysis, abstracting, and indexing. They can also be employed by technical and trade publishing houses and software developers.

The following areas are profiled in this chapter to illustrate the types of positions available in the chemical information field: patent searcher, document analyst, information center manager, textbook editor, scientific translator, and business/competitive intelligence analyst. The profiled careers in this chapter are not exhaustive, but it may give you a starting place to begin exploring the field of chemical information.

Patent Searcher

Stuart M. Kaback

Education
BA, MA, and PhD in organic chemistry, Columbia University

Career Path
While seeking a laboratory position at Esso (now Exxon) Research and Engineering Co. after his graduate studies, Stuart M. Kaback was offered a

position with the Information Research group of Esso's Technical Information Division. After initial hesitation, he talked to several of the chemists and engineers in the group and realized that he might enjoy the experience of participating in a think tank whose mission was to maximize the use of available information.

Except for a nine-month stint in a laboratory and several months editing an abstract bulletin, all of Kaback's 36 years have been spent in what is now the Information Research and Analysis Section of Exxon Research and Engineering Co., achieving the rank of Scientific Adviser, the second highest technical position.

While Kaback's responsibilities are technical rather than administrative, he is responsible for technical supervision of all patent studies and supervises all the reports issued by the people doing patent searching. He spends much of his time building a database of patent information that is of interest to Exxon including bibliographic information, a technical precis, related patents, legal status, and commentary by Exxon attorneys. It is searchable not only by patent searchers but also by patent attorneys and research scientists at Exxon.

Kaback has enjoyed his career and he plans to continue working for as long as he can. He has had consulting offers so if and when he tires of what he is currently doing, he will consider consulting.

Over the years Kaback has specialized increasingly in patent information and has become a well-known authority in the area. His counsel is sought by information suppliers and users. He has many publications and presentations, authored an article on patent literature for the *Kirk-Othmer Encyclopedia of Chemical Technology* (4th ed.), and for 13 years wrote a column on on-line patent searching for *World Patent Information*.

Kaback knows the patent literature and maintains a high level of technical capability in the searching of patent literature. His major responsibility during his career has been to prepare critical reviews of the literature, especially the patent literature, in areas of interest to the company on subjects including petroleum refining, nonconventional energy sources, pulp sources for kraft paper, catalysis, and olefin polymers. He is a trained scientist highly respected by laboratory scientists, attorneys in the Patent Department, and by publishers of patent information.

Kaback feels that he has changed his field by his creative approach to the patent literature. He has also been instrumental in getting database producers to adopt his ideas for making the information more accessible so that it can be searched more effectively.

Career Advice
Beyond a broad knowledge of chemistry, Kaback feels that his liberal arts undergraduate education helped with the necessary communication skills. His training in patent literature was developed on the job, and he feels he learned much from his colleagues. Kaback has benefited from networking and still networks extensively with Exxon colleagues and with patent information specialists throughout the world. Kaback always operates on the principle that if something looks strange, or if the answer is unexpected, always follow up to find out why it looks strange!

Document Analyst
Tony G. Hage

Education
BS in educational chemistry, University of Beirut, Lebanon; MS in organic chemistry and PhD in carbohydrate chemistry, University of Lyon, France

Career Path
Tony Hage took a post-doctoral research position on the preparation of antitumor drugs at the State University of New York at Buffalo and a post-doctoral position at the University of New Orleans on highly condensed polycyclic explosives to be used in space. During his second post-doctoral, he taught general and organic chemistry. Hage's extensive education and post-doctoral work experience in carbohydrates, antitumor drugs, and polycyclic materials, and his fluency in three languages (Arabic, French, and English) led him to Chemical Abstract Services (CAS) to do carbohydrate document analysis.

At CAS, a document analyst like Tony Hage reads an article or document, indexes all the compounds, old and new, and enters the structures, concepts, and keywords in the computer. In general, an article takes two hours while patents take longer. Other specialists abstract the documents. Since there are no college courses on document analysis, he also must be prepared to train other employees.

Hage enjoys his work because every day is different; he enjoys the challenge, feels he is growing, and has a sense of constant improvement. He wants to continue to be successful and to work where he is needed. He feels that while he is serving CAS he is serving the scientific community and the world at large.

Career Advice

Hage feels that any degree in chemistry would suffice to be a document analyst at CAS because CAS trains people in document analysis. One must have the flexibility to handle different types of documents because each document is different. A document analyst needs to be able to use the computer for long periods of time, and must like chemistry.

Information Center Manager
Rose Ann Peters

Education
BS in chemistry, Saint Mary's College, Notre Dame (IN); master's in library science, State University of New York at Buffalo (SUNYAB)

Career Path
Rose Ann Peters is the manager of scientific information for Westwood-Squibb Pharmaceuticals, a division of Bristol Myers-Squibb. She had worked in the lab for 11 years when the plant was expanded and a library was added. Peters felt she had reached her full potential in the lab position, so she volunteered to organize the library. Peters was offered the position of supervisor of scientific information and while she didn't know very much about running a library, it was a startup operation and a perfect opportunity. In time, she realized that she needed a skilled person with a master's degree in library science (MLS). Peters hired one and subsequently went back to school to earn an MLS degree herself.

Peters' library is small. It fills a niche specializing in dermatology research and development while other company libraries specialize in other areas. Peters does much of the scientific literature searching and reference work herself and has branched into business searching and reference services. She also is involved in the collection development, selecting books and journals, and manages technical services like ordering and processing. The cataloging work is outsourced.

Peters particularly likes to interact with people. She sees herself staying in the library and continuing to do what she likes to do: helping people use information to do their jobs better.

Career Advice
Peters highly recommends a library degree in addition to a chemistry degree, although it is not mandatory. She feels that it is beneficial to get a

job in a library and then get the degree that is relevant to that work. If your goal is to work in an academic library, then it is necessary to have a BS and MLS before applying for a library position.

Peters feels that beyond technical skills, other important characteristics to have are people skills, inquisitiveness, analysis from a scientific background, and computer expertise. Another important characteristic is to have the self-confidence to work with people who have advanced degrees, and have reached higher organizational levels. As an information center manager, you need to be a teacher in a variety of settings and be able to advertise and market your services.

Textbook Editor
Sandra G. Kiselica

Education
PhD in bio-organic chemistry, Pennsylvania State University

Career Path
Sandra Kiselica wanted to combine her chemistry background with the business of chemistry, to be able to work with chemical educators, and to use her writing skills. While attending an ACS meeting she spoke to a publisher at Saunders College Publishing, a textbook publisher. Saunders was interested in publishing new textbooks on organic chemistry and biochemistry; as a result Kiselica was hired as a developmental editor. She regards herself as being in the right place at the right time.

Kiselica is still employed by Saunders as a freelance senior developmental editor, working out of her home while she takes care of her young children. Since Saunders is located less than an hour from her home, she can go in to the office when necessary; however, most of her work is done by mail, computer, and the telephone.

Kiselica acts as a liaison between publisher and author. Most of the textbook authors she works with are college professors. She receives a hard copy manuscript that includes formulas and artwork, and identifies other college professors who review the manuscript for content and subject treatment. She then summarizes the reviewers' comments, edits the manuscript, and goes back to the author who approves and completes the editing process.

Kiselica works with an author to enhance the marketability of the book by aiming for the appropriate level and organization of the material, and determining pedagogical features (such as chapter objectives, end of chapter summaries, and computerized exercises). Kiselica must be familiar with competing books and the needs of the marketplace and works with the author and the artists and designers at Saunders to develop the appearance of the book. Kiselica also oversees the development of ancillary materials (from study guides and instruction manuals to videos and CD-ROMs), identifies people to prepare these items, manages their activities, and coordinates the process so that everything is published simultaneously.

Kiselica enjoys what she is doing and is happy that she can combine her family life with a professional career, so she intends to continue what she is doing for the foreseeable future.

Career Advice
Kiselica learned publishing on the job. She states that publishing requires skills such as being organized, being able to set and meet goals without close supervision, and having an ability to handle a variety of tasks simultaneously. The job requires flexibility since each book and each author are different. Familiarity with word processing and computer editing skills are necessary. Furthermore, you must know your field, and be a good writer who can communicate with authors and students. Most of all, Kiselica states, you have to be creative to create a book that stands out from its competitors.

If you are interested in a career in scientific publishing, Kiselica advises attending an ACS national meeting. She suggests that you familiarize yourself with the books at the exposition and talk to the publishers about trends in the market, and talk to educators to learn what they want in their classrooms.

Scientific Translator

Cathy Flick

Education
BS in chemistry, Marywood College, Scranton (PA); MA in physics and PhD in chemical physics, Kent State University (OH)

Career Path

Cathy Flick started her career by teaching physics at Earlham College in Richmond, IN. While teaching at Earlham she saw a translation agency ad for scientists to translate scientific articles and started doing freelance translating. She soon decided to translate full-time. For over 16 years, she has been a regular panel translator for the Russian journal translations program of Consultants Bureau Division at Plenum Publishing Corporation. She also receives translation assignments from other sources.

According to Flick, a scientist does not have to make a choice between science and languages because there is opportunity to exercise both in a career as translator. She has always been interested in languages and acquired a reading knowledge of French, Russian, German, Spanish, Italian, and Latin. Besides a general competency in physics, chemistry, and mathematics, she is familiar with the literature in liquid crystals/model membrane studies; magnetic resonance and other spectroscopic methods; coordination chemistry; biophysics, biochemistry, and pharmaceutical chemistry; polymer chemistry/physics; and the physics of metals and semiconductors.

The field of scientific translator also offers some flexibility: Translating can be a full-time or part-time job; there are freelance opportunities as well as in-house part-time research and translation jobs; or you can be a technical librarian and a translator. It is also a job one can do after early or regular retirement.

Flick does not anticipate that scientific translation will be done by machine in the near future, so she plans to continue to translate for years to come.

Career Advice

Flick's advice to students who have an aptitude for languages is to take as much foreign language and science coursework as possible. Flick took German in grade school, French in high school, Russian in college, and taught herself Spanish and Italian. It is common for scientific translators to learn other languages by studying on their own.

It is difficult to obtain training in scientific translation. The American Translators Association does sponsor programs and workshops in translation. A translator must know how to use the library and must do background reading to capture the nuances and jargon of the field. According to Flick, a translator should translate into one's native language.

Flick states that to be a good scientific translator you really need to be a scientist first. A PhD is not a requirement but you do need to have as broad a technical background as possible. Scientific translation is a good job for anyone who is interested in languages, has a reading knowledge of the source languages in the relevant field, good writing skills in the target language, persistence, attention to detail, a good general background in the relevant fields, and a knack for digging out obscure information on a wide variety of subjects in the library or elsewhere. Flick says most translators like to work alone when actually translating; she describes a translator as a happy hermit.

Business/Competitive Intelligence Analyst
Myra Soroczak

Educator
BS in chemistry and mathematics, Florence State University (now the University of Northern Alabama); MSE in materials science, Louisiana State University.

Career Path
Myra Soroczak works in the Tennessee Valley Authority (TVA) Environmental Research Center, which looks at developing environmental technologies that are of value in the Tennessee River valley and in the nation. The TVA is an independent federal corporation that manages the dams and water of the Tennessee River for flood control, power generation, recreation, and industrial uses.

Soroczak worked as a chemical lab analyst at TVA for three years and as an associate engineer at IBM for three years; she then returned to TVA as a research chemist. After 12 years, she obtained a master of engineering degree in materials science and became a research manager.

Soroczak describes her job as providing management with information and analysis in order to make business decisions. Because she has always been an avid reader of all kinds of information and has been interested in keeping up with what is happening in her field, she seized the opportunity to become a specialist in the business intelligence area. As an information analyst, she relies heavily on secondary sources like newsletters, trade publications, and company annual reports and on searching the Internet and Dialog. Occasionally the information must be verified and validated by speaking to someone at the source of the information.

Competitive information is a constantly moving target. It is necessary to keep current on in-house developments in order to assimilate useful and relevant information from the outside. To serve the client properly, Soroczak must determine how to format the information, and decide whether the client wants detail or just the bottom line.

Soroczak likes what she is doing and would like to stay in business intelligence. She is contemplating obtaining an MBA to help her understand the business and financial parts of the intelligence, to make herself more valuable, and to be in a position to shift to market analysis if competitive intelligence goes out of vogue.

Career Advice

In addition to a technical background and computer skills Soroczak feels that an ability to spot trends is vital. She knows of no training specific to developing this skill, but feels that any training in logic is useful. Communication and people skills are also very important. A business intelligence specialist must be able to work alone and must be intensely curious—but also knows when to stop asking why. There is always more information than the client wants, so the specialist must be able to sift out the important facts. Soroczak states that working as business intelligence specialist is interesting but is not for everyone because one's credibility is always being tested.

— 5 —

Computer-Related Careers

Computer applications affect every field of chemistry and every chemist should be computer literate. However, there are some specialized areas of computer applications that go beyond basic computer literacy that may appeal to chemists.

In general, the skills required in this field are a comprehensive knowledge and use of computer procedures including programming, often in fourth generation, object-oriented languages. The computer scientist must be creative, understand customer needs, be good at problem-solving, and be able to apply statistics. The computer scientist must have good communication skills, including writing, presenting, listening, and convincing. Because a computer scientist rarely works alone, the ability to work as part of a multidisciplinary team is also essential.

There are as many kinds of computational chemists as there are fields of chemistry. Some areas in which computational chemistry has played an important role are: the design of pharmaceuticals, polymers, materials, and agrochemicals; analytical chemistry/laboratory automation; data analysis; spectroscopy; and most recently, combinatorial chemistry.

This chapter contains a few examples of how some chemists went into computer-related careers. The following areas are profiled: computer modeling, process automation and control, software development, and software evaluation and documentation.

Skills and Characteristics

- Computer skills
- Programming ability, often in fourth generation, object-oriented languages
- Creativity
- Communication skills, oral and written, especially technical writing
- Customer-service skills
- Problem-solving skills
- Listening skills
- Ability to use statistics
- Presentation skills

Computer Modeling

M. Katharine Holloway

Education
BS in chemistry, University of Southern Mississippi; MS and PhD in theoretical organic (computational) chemistry, University of Texas at Austin

Career Path
M. Katharine Holloway does molecular modeling of potential pharmaceuticals at Merck Research Laboratories. During her graduate coursework with Michael Dewar at the University of Texas at Austin, she started working with liquid crystals and did some computational chemistry to support her research. Holloway states that she "got hooked," and never went back to liquid crystals. She went directly from graduate school to Merck where she performs molecular modeling studies using 3-D molecular graphics to examine the interactions of small molecules such as drugs with macromolecules like enzymes or receptors. She was recently involved in the development of a new AIDS therapy, CRIXIVAN™, which is an HIV-1 protease inhibitor.

Holloway works in an applications-oriented group and is assigned to a project team supporting ongoing drug development using commercial or in-house developed software tools. She tries to understand and predict the relationship between structure and activity in order to design better drugs. She also employs 3-D information to perform database searches to identify new lead structures.

Holloway sees new vistas for computational chemistry such as combinatorial chemistry and three-dimensional quantitative structure-activity relationships. She is happy performing and publishing scientific research and has opportunities for advancement on the scientific track. She is currently working on integrin receptor ligands for the prevention of osteoporosis and inflammation; cardiotonics; and antibiotics.

Career Advice
Most senior people doing computer modeling have advanced degrees in chemistry, biochemistry, physics, or biophysics. Support personnel usually have either a chemistry background and learn about computers, or have a computer science background and learn about chemistry.

Beyond the technical expertise, problem-solving and communication and people skills are essential. Holloway works on multidisciplinary teams so she must be able to communicate what she is doing, listen for questions, and identify critical elements that will help move the project forward. For instance, a team may consist of medicinal chemists, biochemists, biologists, pharmacologists, X-ray crystallographers, and NMR spectroscopists, as well as computational chemists.

Holloway feels that in addition to the knowledge of chemistry and computers, a computational chemist can benefit from expertise in another field such as medicinal chemistry or biochemistry in the design of pharmaceuticals, or macromolecular structure and electronic structure theory in the design of polymers and materials.

Process Automation and Control
Joseph L. Maglaty

Education
BS in chemistry, Providence College, Providence (RI); PhD in analytical chemistry, Colorado State University, Fort Collins

Career Path
Joseph L. Maglaty obtained a PhD in analytical chemistry specializing in data acquisition, computer automation, and interfacing acquisition sys-

tems with spectroscopic instruments. He is at Merck and Company, Inc. in the Research and Information Systems, which is dedicated to systems development for the Merck Research Labs. He started at Merck as a project manager for laboratory automation and has been there for 13 years. His work runs the gamut from writing small programs on laptops for data acquisition from laboratory instruments to a multimillion-dollar team project for a clinical area using client server systems with an isolated database server. The information is retrieved from a number of instruments, processed, sorted, and analyzed in a laboratory data management system which collects and processes data and generates worksheets in support of vaccine clinical trials.

Maglaty's title is business analyst, which means that he is the liaison between a Merck business area and the information technology organization. He must understand the business and its objectives, then work with a research area to fill its information technology needs. His responsibilities include project management, project definition, staffing, and staff management to support the products after they have been established as products or potential products.

Since he is in a project management position, Maglaty supervises the people on the team. Maglaty's next step is to a mid- or upper-level management position to oversee projects in several areas of the business to support the research environment. A move into corporate management is also a possibility.

Career Advice

Maglaty says that in his capacity a background in life science, not necessarily chemistry, is needed in addition to computer science training and background. He recommends taking courses that involve programming in the newer, fourth generation object-oriented languages as well as business courses. Communication skills, especially technical writing skills, are necessary to write the necessary project documentation. An ability to manage projects is also important.

Maglaty points out that he enjoys working with people, and that team-building skills are important in his job. He has gone beyond sitting in front of a computer; he works on developing and managing a project. He feels that all pharmaceutical companies rely heavily on information technology and are looking for people with his combination of skills.

Software Development

Neil Ostlund

Education
BA in chemistry and physics, University of Saskatchewan; MSc and PhD degrees, Carnegie-Mellon University

Career Path
After a postdoctoral fellowship at Harvard, Neil Ostlund rose to Associate Professor with tenure in the Department of Chemistry at the University of Arkansas. He was a Visiting Associate Professor in the Department of Computer Science at Carnegie-Mellon University and then was an Associate Professor in the Department of Computer Science at the University of Waterloo. He spent a summer as a systems analyst at the Saskatchewan Power Corporation before graduate school and as a special consultant on supercomputers at Intel Corporation while at the University of Waterloo. His experience in industry taught him that he wanted the freedom of teaching and working for himself rather than for a company.

Ostlund was not satisfied just to teach chemistry. He saw that chemists who were working in molecular modeling had needs which the software could not support, so he developed software with an open, flexible architecture to help address their needs. He is the founder and CEO of Hypercube Inc. and the developer of Hyperchem®. He has been writing software since college and while he does not personally write software now, he does write the specifications for products like Hyperchem®. Ostlund currently prefers the research environment and is satisfied with his career, but thinks it is possible that he may become restless in the future and look for another career.

Career Advice
Ostlund advises someone who is interested in programming and writing software to get a PhD and then find an interesting place to work. Ostlund feels that beyond a background in chemistry and computer science, further training is unnecessary. He feels that his training in chemistry and in computer science gave him the tools to program and to develop software, as well as to found and run a company. Ostlund also emphasizes the need for excellent writing and communication skills, as well as problem-solving skills.

Software Evaluation and Documentation

Lisa M. Balbes

Education
BA in chemistry and psychology, Washington University, St Louis (MO); PhD in organic chemistry, University of North Carolina at Chapel Hill

Career Path
Lisa M. Balbes is a scientific software consultant who provides support services such as software testing and documentation evaluation, and writing and editing to scientific (primarily chemistry) software developers who want to improve the quality of their software. She uses Macintosh, IBM DOS, Windows 3.1, Windows 95, Unix, and VMS platforms.

While in graduate school Balbes did some computer modeling using commercial software. She decided not to continue in lab work in part because she wanted to have a family, so she took a post-doctoral position and worked at the Research Triangle Institute for just under three years doing computer modeling.

Both she and her husband, also a PhD, found employment in Columbus, Ohio. After they moved to Columbus her position at a supercomputer center was not funded so she found herself back on the job market. About that same time, one of the companies whose software she had used in her research called to offer her an opportunity to evaluate some new software before its public release and offered to pay for the evaluation. It then occurred to her that there might be other companies who needed her services, so she set up her business at home.

Balbes tests the software as if she were the user and makes sure that the documentation does indeed tell the user in clear language how to use the software. She not only edits but extensively rewrites the documentation if necessary. There are over 20 organizations and publications who have used her services, including Tripos and John Wiley & Sons.

Balbes enjoys her work and intends to continue. She may branch out into writing documentation, particularly translating technical information for the layperson. More companies are outsourcing documentation writing and software testing. She feels that she has a great job because it is portable enough to do not only from home but even while traveling.

Career Advice
Balbes states that software testing is a growing field; beyond the chemistry degree she recommends computer science courses, business courses, and

experience with software packages. She wishes that she had worked longer with software before starting her business, not just for the software experience but also for better networking opportunities.

The ability to be critical and detail-oriented is essential. Self-motivation is very important since there usually are deadlines to be met, and in her independent capacity, Balbes must be alert to new business opportunities. Interpersonal skills are also important. Balbes advises experience with a variety of software packages; quality assurance training; and an established network of software developers. An additional requirement is having the latest in computers and programs.

—6—
Conservation of Art and Historic Works

Conservation of art objects such as paintings, sculpture, and furniture, and of historic works, such as letters, notes, manuscripts and diaries, and rare books, requires the skills of conservators and conservation chemists. Unless one happens to have a strong background or interest in art or history, a career in the conservation of art or historic works may not immediately come to mind.

Conservation is primarily concerned with preservation. First, a conservator examines and documents the condition of an object or work. Second, if the conservator advises restoration or repair, a treatment plan is devised and implemented. Third, whether the object is treated or not, its preservation is undertaken.

A fourth area of practice is called scientific analysis and research. Here a conservation chemist, or other materials scientist, applies a variety of analytical tools to learn about the materials used in the object of interest. This information may help to date the piece of work, and with the chemist's knowledge, help the conservator to decide on a course of treatment and preservation. The chemist also does some conservation research such as improving preservation processes and materials, or improving analytical techniques.

Conservation is part science and part craft. For example, the restoration of furniture requires wood working and finishing skills. Manual dex-

> ## Skills and Characteristics
>
> - Extensive knowledge, interest, and experience in an art, or arts, or in books and libraries
> - Manual dexterity, necessary for the conservator and handy for the chemist
> - One or more craft skills (such as painting, finishing, woodworking, pottery) necessary for a conservator and helpful to a chemist
> - Patience; careful work can be tedious
> - Decision-making skills, especially for a conservator and helpful for a chemist
> - Communication skills, oral and written
> - Interpersonal relations
> - Instrumental analysis and analytical chemistry, to be a conservation chemist
> - Materials science knowledge, necessary for a conservation chemist, and helpful to a conservator
> - Computer skills
> - Persistence and creativity

terity is a necessity. Many of the repairs are in very small areas of a work, such as in a painting. A slip of the knife may not only cause personal injury, but could also damage an art object and significantly lower its value. Even the conservation chemist needs a level of manual dexterity to handle and analyze the very tiny, precious sample allowed to be taken from a valuable work.

A chemist who is interested in a career as a conservator should have extensive knowledge and interest in the subject works and objects, craft abilities related to these subjects, and manual dexterity. For a conservation chemist, additional skills include instrumental analysis and materials science, and craft capabilities can be helpful at times. An example of the latter is if you paint oils or watercolors for recreation, you will have a greater feel for a conservator's challenges in working with paintings.

Conservators and conservation chemists work in the larger, better-funded museums, institutes, and libraries. The number of positions is limited, and the competition is keen. A conservation chemist requires a degree in chemistry, so all other things being equal, a chemist has training that could enhance his or her competitiveness for a conservator's job. A source of information about conservation and careers in conservation is the American Institute for Conservation of Artistic and Historic Works, 1717 K Street, Suite 301, Washington, DC 20006.

Increased insights into careers in conservation can be gained from the career paths and information supplied in the following two profiles.

Conservation Chemist
Beth A. Price

Education
BA in art history, BS in chemistry, State University of New York (SUNY) at New Paltz

Career Path
Beth Price is a Conservation Chemist with the Philadelphia Museum of Art. Her interest and degree in chemistry came after her first interest and degree in art.

Price started college with an interest in studio art. After she received her BA in art history she taught art, worked in a potter's studio, and then decided to study biomedical illustration in graduate school, which required science courses. Price took chemistry and found to her surprise that she liked it. She had an excellent general chemistry instructor and also enjoyed organic chemistry. She received her BS in chemistry and was encouraged to go to graduate school in chemistry, but she decided to seek a job.

Price worked as a research assistant in the SUNY Research Foundation for one year, and then in the analytical laboratory of the New York Botanical Garden Institute of Ecosystem Studies in Mill Brook, NY for a summer. In 1984, Price joined the FMC Corporation Agricultural Chemical Group at the R&D center in Princeton, NJ. There she synthesized, isolated, and determined the structure of biologically active compounds using analytical instrumentation methods. She took both evening courses and short courses related to chemistry, as well as art classes.

In 1990, Price answered an advertisement in *Chemical & Engineering News (C&EN)* for a two-year fellowship as scientist at the Harvard University Art Museum. She interviewed and received an offer. Several months earlier, the Philadelphia Museum of Art (PMA) had advertised for a conservation chemist but since the ad stated that a PhD was required, Price had not applied. In order to find out what a reasonable salary was for the Harvard job, she called PMA to ask about the salary for the previously advertised position. She was told that their position was still open and they were interested in interviewing her. Price was offered the job. She notes that the combination of art history and chemistry was an important factor in getting the interviews and offers at both places. She accepted the PMA offer and became a Conservation Chemist at the Philadelphia Museum of Art in 1990.

Price's role as a conservation chemist is to support the work of the conservators in maintaining, repairing, and restoring works of art. At first one may think only of working with great paintings, but there are also picture frames, furniture, tapestries, iron works, sculptures, vases, clothing, castings, carvings, glass, ceramics, woods, metals, plastics, finishes, coatings, fibers, fabrics—a different world of materials and material science associated with a major art museum. She analyzes the materials in the artwork to help the conservators restore or stabilize the piece. With her chemistry knowledge, she helps them plan their treatment strategy.

Price also helps the museum evaluate paintings that it is interested in purchasing. Analytically determining the kind of pigments can help to date the piece. Sample sizes are often no larger than one milligram.

In cooperation with conservators, she also does materials studies, such as application techniques or aging studies using coated coupons. Work is also done for other public institutions which do not have analytical equipment. While the conservator's first responsibility is to PMA, some of the grants received for conservation stipulate helping other institutions lacking the PMA's capabilities. Price states that there are only 15 museums in the US that have analytical chemistry facilities associated with them. The Smithsonian Analytical Lab outside Washington and the Getty Conservation Institute in California are two of the largest.

Price feels that while she has settled into her job, the work is still challenging and stimulating because there are so many materials with which she works, and because there is always something new to learn. She really enjoys the combination of art and chemistry, and in spite of the limited opportunities, she would encourage anyone who loves both art and chemistry to pursue a career as a conservation chemist.

Career Advice

A degree in chemistry is required for a job as conservator, with an emphasis on analytical chemistry and instrumentation. It is no surprise that a strong materials background is important. Computer skills are necessary to do word processing, draw chemical structures, search the literature, and maintain databases. Price created one database as a sample reference library related to conservation for the fluorescence microscopy and FT-IR spectroscopy instrument. Like academe, publications in journals and presentations outside the institution are necessary.

Good communication and interpersonal skills are also needed. Price is involved in raising funds for analytical lab equipment, often talks to other scientific experts, and takes samples to other laboratories for analysis. She also works with the conservators, explaining the chemistry to them. Diplomacy is required when conservators have staked their reputations on characterizing a piece of art before the chemical analysis and the analysis does not support the characterization. Price is often called on to give tours of the Conservation Department and show or demonstrate the lab equipment.

Price advises those who are interested in becoming a conservation chemist in the museum community to recognize that the number of opportunities are limited. Pursue a graduate degree in materials science, and learn about art, art materials, and the aging of materials. Become very knowledgeable about the computer, analytical instrumentation, and chemistry.

Conservator

Whitney S. Baker

Education
BA in chemistry and Spanish, University of Kansas; pursuing a Master's in information and library science (MLS) with an advanced certification in conservation, University of Texas at Austin

Career Path
As an undergraduate, Whitney Baker worked in the library at the University of Kansas retrieving books from the stacks for interlibrary loans. She was appalled by the condition of the books. A professor she spoke to about possible career paths mentioned the work of library conservator and suggested that Baker get further information. This led to an internship at the Center for the Book at the University of Iowa in Iowa City.

Baker worked with a paper maker associated with the Center for the Book. The paper maker prepared special papers on a small production scale and did research on papers. Baker took courses in paper making, non-adhesive bookbinding, and library science at Iowa. She also took the opportunity to work in the conservation lab associated with the Center for the Book.

One objective for Baker at Iowa was to find out more about conservation work itself and determine her interest in it. Another objective was to build a portfolio showing her manual dexterity and interest in the field of library conservation, a requirement for admission to the MLS program at the University of Texas. Baker found that her interest in conservation combined her background in organic chemistry with her interest in protecting and restoring books, and she calls the work a combination of science and craft.

Baker started her three-year MLS program at the University of Texas-Austin, pursuing a master's in information and library science (MLS) with an advanced certification in conservation, in September 1995. The program is dedicated to Library Conservation. There are other schools for conservation, but they are more art-focused.

Baker explains that there are four areas of conservation: examination, treatment, preservation, and scientific analysis. First, the book or collection of books is examined and their condition documented with notes and photos. The objective is to determine the extent of the damage and deterioration and to decide if it can be repaired or restored and how. Sometimes it is decided that it is better not to try to repair, but to move on to preservation.

The second area is treatment or repair of the book or collection. The repair or restoration has to be reversible; it is part of the code of ethics for conservators. The third area is the preservation of the books to retard deterioration; sometimes books are put in special boxes or in protective wrappers. Sometimes paper is encapsulated with clear, inert sheets of Mylar® polyester film. The object may not have been treated, but further degradation is inhibited. Finally, the scientific analysis and research area is usually left to conservation scientists like Beth Price. Many times there are no funds or facilities to do scientific studies in library conservation.

After graduating with her master's, Baker plans to do an internship on the east coast where there are a number of libraries with conservation departments, followed by one or more post-graduate fellowships. Her ultimate goal is to work as a conservator in a library and do research. Baker would also like to collaborate, using her Spanish, with conservators in Latin America and South America, where there are many problems facing library conservators, including the climate, politics, and economics.

Career Advice

To pursue a career in library conservation, it is best to follow a career path like Baker's. Get a job in a library, even if it is as a volunteer. Make contacts with people who work in conservation, write letters, follow up, ask questions, and gather information. Volunteer to work on projects that involve interacting with a conservator; you can earn respect for volunteering, as well as knowledge, experience, and contacts in conservation. Two journals associated with conservation are the *Journal of the American Institute of Conservation*, and for library conservation, the *Paper Conservator,* published in the United Kingdom.

The primary educational requirement for a conservator in library science is the master's program at the University of Texas, presently the only program in the US. There is also a post-graduate, four-year apprentice program for a limited number of people at the University of Iowa. Baker added that there are other conservator programs for works of art at New York University, the University of Buffalo, the University of Delaware-Winterthur Museum, Harvard University, and Queens University in Kingston, Ontario.

While Baker has an undergraduate degree in chemistry, many of her classmates have degrees in art. To be admitted into the Texas program, applicants must have two years of college chemistry, including organic. The curriculum in the master's program stresses chemistry and includes graduate chemistry classes.

Other requirements and characteristics important for a library conservationist include manual dexterity and patience. Communication skills are important, both written (for documenting work and publications) and oral (to communicate with other people on the staff or at conferences). Decision-making skills are necessary since the process of working with a book or collection may not be clear, yet decisions must be made. Since many institutions can only afford one conservator on staff, working independently and alone is often required.

—7—
Consulting

Consulting is generally considered a field for experienced people with a long and/or broad range of experience in the chemical and chemical-related industry. Consulting has become a significant career alternative for chemists, and can offer career alternatives at any stage of a career.[1]

There are a variety of ways to enter the consulting career path. Some chemists combine their technical training with a non-technical background, such as business, and work for an established consulting firm early in their careers. Some early career consultants distinguish themselves with a technical specialty, especially in an emerging technology or market.

Some chemists work under contract as temporary employees, and a typical assignment is in the lab, but an assignment could be in other technical-related activities such as literature or patent searching, market research, or customer service. Because these contract workers are referred to as consultants, they have been included in this section.

Many chemists close out their careers as independent consultants. These chemists have a reservoir of experience, mostly with large chemical, petrochemical, petroleum, or pharmaceutical companies. They may also have broad experience including management, marketing, product management, manufacturing, sales, and patents. Some chemists choose to consult because they want to be more independent. Others choose consulting because they have lost their job or were retired earlier than they had planned. Finally, there are those who retire but want to continue working professionally on their own terms.

---

[1] *Chemistry & Engineering News,* February 26, 1996, p. 43, reported that the 1995 Salary and Comprehensive Survey found that an estimated 27,000 ACS members appear to be engaged in some kind of consulting work.

> ## Summary of Skills and Characteristics
>
> - Chemistry or other strong technical degree and background
> - Business knowledge and judgment
> - Communication skills, both oral and written
> - Problem-solving skills
> - Marketing and networking skills
> - Analytical skills and the ability to work with a minimum of information and data
> - Good interpersonal skills
> - The ability to work on a team
> - Flexibility
> - High level of self-confidence
> - Computer skills at a level required for your particular business or specialty
> - Continuous learning, reading
> - Persistence

Consultants seek out assignments, establish client needs, write project proposals and reports, extract and analyze information, establish supported conclusions and recommendations, and make presentations. There is a large business component to consulting: Marketing, networking, billing, accounting, and advertising are needed to support a consulting business. Some consultants publish bimonthly newsletters as part of their marketing effort not only to inform potential clients about topics which may be of interest, but also to help keep the consultant's name and specialty in mind.

There are no formal requirements to be a consultant. A technical degree like chemistry is necessary, but not sufficient, to be a consultant. Understanding business and having business judgment are absolutely necessary. Science and engineering consultants can have BS, MS, or PhD degrees. While a PhD can add status to the business card and stationery, client satisfaction and word of mouth advertising and recommendations are more important.

Computer skills are also necessary. Word processing ability is needed to generate letters, faxes, proposals, bookkeeping, and reports. Spreadsheets, particularly for certain technical, cost, financial, or business-related consulting may be necessary. Using presentation software to prepare high-quality overheads can be helpful. Beyond these needs, other computer skills depend on a specific consulting area.

An independent consultant must be prepared to handle administrative tasks that may ordinarily be handled by a secretary such as typing letters and reports, buying supplies, answering the phone, sending faxes, and getting the computer repaired.

Most of the consultants interviewed in this chapter agreed on a number of important personal skills and characteristics. Strong interpersonal relations and communication skills, both oral and written, are critical. Brevity and clarity in writing are valued. Analytical ability, problem-solving skills, and flexibility are important. Marketing and networking are required for success. The ability to obtain information from various sources and identifying the pertinent information is important. A high level of self-confidence is required since the rejection rate of proposals is high.

Consultants can be helpful to larger organizations that are outsourcing work previously done in-house. Likewise, smaller companies may lack many of the resources of the larger companies and hire consultants to gain the benefit of their technical and business knowledge and experience.

Sometimes contract chemists have a strong background and record of accomplishment in specific technical areas where formulation is key. These areas of expertise could range from cosmetics to adhesives to elastomers. Others may bring experience in instrumental analysis or synthesis. Many times these temporary jobs lead to permanent jobs for the chemist.

There are several ways to find contract work. Classified ads in newspapers carry advertisements from job placement firms specializing in recruiting scientific and engineering personnel for temporary jobs. The Yellow Pages® can be a source of finding temporary placement services for chemists. Networking and asking other chemists whether they know of any temporary professional openings or recruiting firms is another route. Even talking with a knowledgeable person at a temporary employment office might uncover information on what companies are sources for temporary jobs for chemists. Other possible sources for temporary work are through consulting organizations like The Chemists Group.

Most consultants are independent, but they may be associated with consulting groups or networks to enlarge their marketing efforts and assignment opportunities. Sometimes an independent consultant's business card may read "Smith & Associates," implying a group of several individuals. Many times there are no real associates as a part of the defined business, but the consultant may know others to whom they could refer clients when the requested service is not a specialty of the consultant. Some clients prefer to work with a multi-consultant organization in the belief that consultant availability and response time will be better, or associates could take over a project in the event the consultant cannot complete an assignment.

Whether a chemist is an independent consultant or is looking for temporary work, associating with larger consulting networks or groups can expand opportunities. Some examples are: The Technology Group in Stamford, CT; the Cecon Group or Condux in Wilmington, DE; or Omnitech in Midland, MI. One source of general information on consulting organizations is *Consultants and Consulting Organization Directory,* edited by Janice McLean and published by Gale Research Company (1992).

Choosing a local consulting group could maximize chances for temporary work and be convenient for the independent consultant. The Yellow Pages® or the "Business to Business" telephone directory lists some consulting organizations. Participation in professional organizations, such as the American Chemical Society, and formal technology and consultant networks can also open career opportunities.

This chapter contains examples of how some chemists went into consulting. It is important to bear in mind that the career paths that these individuals have taken are unique; there are as many approaches to becoming a consultant as there are consultants. These profiles may give you a better understanding of the possibilities.

Independent Consulting

Geoffrey Dolbear

Education
BS in chemistry, the University of California at Berkeley; PhD in chemistry, Stanford University

Career Path
Dolbear founded G. E. Dolbear & Associates in Diamond Bar, CA in 1989, after working 24 years in industry. After receiving his PhD in chemistry from Stanford in 1966, he worked for DuPont, W. R. Grace, Occidental

Petroleum, and UNOCAL. In 1969, while at W. R. Grace, he became acquainted with consultants Gene Schwarzenbeck and Ike Yen, whose examples showed him that the independent lifestyle was appealing. Dolbear traces his decision to become a consultant to an August day in 1987, while fly fishing in the Frying Pan River. He felt that his career at UNOCAL, and perhaps in industry, had peaked and that there were limited opportunities for significant growth. He decided to try consulting on his own.

After making his decision, Dolbear began to consult in the evenings and on weekends. Since he was still working at UNOCAL, he avoided any conflict of interest by not consulting in petroleum refining. He also began meeting with Ike Yen for lunch once a month during this period. He describes Ike as his mentor in making his transition to independent consulting. Yen had begun consulting in 1982 in the safety and environmental arena when both Dolbear and Yen were laid off from Occidental. By 1989, Dolbear had more business than he could handle evenings and weekends and left his industrial job to pursue his consulting business full time. He is very pleased with his decision and his experience as a consultant, from both a business and personal standpoint.

Career Advice
Dolbear emphasizes that consultants need to be people-oriented because it is necessary to get along with a wide range of clients and personalities. As a consultant you have to market yourself strongly and often. He suggests not accepting a project that is not compatible with your skills. If you are faced with this situation, you need to recognize this and connect the client with a consultant whose strengths and experience are a good match. This strategy will avoid problems and strengthen your network at the same time. He reads a lot to keep current, particularly about the technology and business area in which he works. He also points out the need to find ways to use his time effectively when he is not with clients. Discipline is important since you are your own boss; setting a schedule and keeping to it can be much harder than when you are working for someone else. Finally, he notes that emotional support from family is important.

Independent Consulting
Don Berets

Education
AB, MA, and PhD from Harvard University

Career Path

Don Berets was cofounder and former President of The Technology Group[2] in Stamford, CT, a firm that matches clients and their needs with consultants based on their area of expertise. Other similar groups are the Cecon Group or Condux in Wilmington, DE, or Omnitech in Midland, MI. The business collects a percentage of the fee charged to the client to cover costs and return a profit. Independent consultants are associates of the group while temporary chemists hired from The Technology Group are employees of the group. Temporary chemists are hired to do temporary work under contract for extended periods, usually in the lab.

After a one-year research fellowship at MIT at the conclusion of his studies at Harvard, Berets joined American Cyanamid and held various research and research management positions, the last being Manager of Government Contracts. In 1986, after a 37-year career, he took early retirement but Don wanted to keep active professionally. While he thought he wanted to become a consultant, he saw a need, along with several others, to set up a company to market other consultants. Initially most of the consultants were from Cyanamid, but soon there were chemists from other companies who wanted to be associated with the group. He took on the administrative work and became President. His role included managing the consultant database, tracking résumés, clients, and consultants, and doing the accounting and payroll duties. He visited clients, published a newsletter, wrote articles for trade publications as a part of the marketing effort, and interviewed new consultants who applied to join the group, as well as performing other assorted office chores.

Career Advice

Berets reiterates the need for business understanding and judgment. The effort to land a client has to produce the appropriate income. He also cites the need for a strong ego since the rejection rate is high: Berets estimates that the batting average is around .250, which is not bad for baseball perhaps, but not great for a consultant's ego. A thorough understanding of marketing is critical. Conflict resolution is an ability that is sometimes needed. Berets urges all chemists to think about their career seriously and consider the breadth of their capabilities and develop them appropriately.

---

[2] Prior to 1996, the group was known as The Chemists Group. In a business change the name was changed to The Technology Group and Berets has retained The Chemists Group.

Consultant in a Large Consulting Firm
Margie Graves

Education
BS in chemistry, MBA, University of Virginia

Career Path
Margie Graves is a consultant with a firm in Alexandria, VA. She enjoyed her undergraduate research in chemistry, and the open discussion and interaction among students and professors that is associated with an academic research environment. After obtaining her BS degree in 1979, she took a job at the FBI crime lab in Washington, DC. She worked in a neutron activation analysis lab three floors underground. The nature of the work and isolated work environment removed Graves from the interaction she had enjoyed during her college research. This isolation, combined with the routine nature of the job, persuaded her to look for a different career path.

Graves took several graduate school exams including the GRE, MCAT, LSAT, and GMAT. She began to rule out certain career paths: she did not want to teach; industrial chemical jobs tended to be in locations that were not attractive to her; law was not appealing. Since she had scored 98% in business in the GMAT, she obtained her MBA from the University of Virginia in 1982.

At the conclusion of her studies, Graves answered an advertisement in the DC area for an entry-level position with a consulting company specializing in technology applications. Two years later she moved to a similar company which did work mostly for the government. After five years there, Graves was laid off when the company was taken over and downsized. She was a self-employed consultant for two years but also used the time to plan carefully for her next career move. In 1991, she joined A. T. Kearney where she was recently promoted to Director of Operations, North America.

Graves has consulted in the areas of strategic planning, market analysis, management systems analysis, and financial analysis. She has experience in business process reengineering, merger and acquisition analysis and execution, organizational assessments, operational restructuring, pricing strategy development, and business plan development for new ventures. Her clients have included public utilities, biomedical firms, defense contractors, information systems hardware suppliers, and federal agencies.

Most of Graves' clients operate in a regulated environment so her chemistry background is important to understanding the environmental regulations and technical issues that affect them. She has also been able to use her analytical and problem-solving skills. Graves' business schooling and subsequent experience are important since all the projects have a considerable business component. She enjoys the teamwork required by her job, which meets her goal to interact with people on a professional basis.

Career Advice
Graves emphasizes that consultants have to deal with a quantity of information, and need to be able to grasp important and pertinent items quickly. She refers to this as "filling the funnel," which takes practice. The ability to communicate with all of the client's management levels is necessary. The ability to answer questions on your feet is important, particularly in presentations.

It is necessary to become politically astute about the client organization and recognize who may be affected by the project's results. Successful implementation of recommendations resulting from a consulting study requires that key members of the client organization be "on board" and convinced that the solution is best for the company.

Graves recommends having a broad education and knowledge, while developing expertise in several subject areas and industries. Continuous learning is important. Finally, she emphasizes the need for self-confidence because as she says, "consulting is not for the faint-of-heart."

Consultant in a Large Consulting Firm
Allen Merriman

Education
BS in chemical engineering, and master's of science in industrial administration (MSIA), Purdue University

Career Path
Allen Merriman is employed at the same firm as Margie Graves. Unlike her, Merriman joined the organization just after graduation, with only a summer of industrial experience under his belt.

After receiving his MS, Merriman sought an entry-level chemical engineering job doing process or project work at a chemical plant. He gained some plant experience after his BS degree from a summer job in a UNOCAL adhesives plant. However, the job market was poor in 1986 and some potential employers considered him overqualified for a plant job because of his MSIA.

Merriman broadened his job search and responded to an advertisement in *The Wall Street Journal* for an opening that combined chemical training and economics. He was hired by the consulting firm, A. T. Kearney, where he is currently a Management Consultant.

In his consulting work, Merriman uses both his technical background and business-related education from his MSIA. Many of his assignments have been in the area of environmental health and safety (EH&S). Recommendations require written and oral reports so the ability to speak and write is important. While most new business is generated by word of mouth and by management, he also spends approximately 10%–15% of his time marketing, networking, and seeking projects for the company.

Career Advice
Merriman feels that some tolerance for risk is an important characteristic to be a consultant. In 1991, an opportunity arose for him to work in the finance area related to transfer pricing on imported products. Although not familiar with this area, he undertook the assignment and successfully completed it. This led to other assignments in management consulting, which broadened his experience and value to the company.

Merriman has a knowledge of plant operations and environmental regulations and technology. While not a requirement for the position, he has become a Certified Management Consultant. He believes that an MBA, or in his case an MSIA, is helpful but not necessarily a requirement; what is required is business knowledge.

Merriman said that it is important to find a job that is enjoyable, interesting, and challenging; such an environment helps an employee find or create opportunities. He also stated that a consultant should market what he or she is doing and can do, both internally and externally, as well as understand what others are doing. This helps to build an internal and an external network.

—8—

Education

The requirements for a career in university and college teaching are generally well understood by most chemists. College teaching, along with industrial laboratory and management positions, is considered a typical career for chemists.

However, there are careers related to education which are outside of the teaching and research arena. Moving from teaching to administrative positions can be a career path. Teachers who choose this route often see it as an opportunity to influence the education of students through mentoring and curriculum development or improvement. Leadership skills are needed to perform effectively as an administrator.

The alternative careers discussed in this chapter are representative of some of the career possibilities in academic administration: secondary science education administration, high school chemistry teaching, high school principal, college and university academic administration, and environmental health and safety in academe.

Secondary Science Education Administration

James Stockton

Education
BS and MS in chemistry, University of North Texas, Denton

Career Path
After completing his undergraduate studies, Stockton planned to go to Officers Candidate School (OCS) to pursue a military career. During a lengthy delay in processing his OCS paperwork, Stockton was asked to

Skills and Characteristics

Teaching High School Chemistry and Science Education Administration

- A degree in the subject taught; a degree in chemistry to teach chemistry in the high schools is increasingly important
- Meet the state requirements for high school teachers or administrators
- A strong desire to teach, and a dedication to the profession
- Strong communication skills: oral, written, and listening
- Keeping current in chemistry and related sciences
- Problem-solving skills
- Decision-making skills
- For administration: Leadership skills, enhanced listening skills, and conflict resolution skills

Academic Administration

- A PhD is almost always required for teaching full-time on the faculty of a university or college
- Teaching and research capabilities
- A prior career as a faculty member achieving tenure and rank of professor is preferable
- Leadership skills: vision, sharing, commitment, decision, execution, credibility
- Excellent interpersonal relations, must like interacting with people
- Good communication skills, oral and written
- Ability to listen
- Conflict resolution skills
- Problem-solving skills
- Exposure to and a record of contribution and accomplishment in professionally related special assignments, committees, boards, teams, etc. within and outside of the university, the latter especially at a national level with ACS, NSF, NIH, etc.

Academic Environmental Health and Safety

- A degree in chemistry and appropriate training, education, and certification in environmental protection, health and safety areas. An alternative is a degree related to environmental health and safety.
- A good understanding of the health, safety, and environmental laws and regulations applicable to university operations
- Effective communication skills, oral and written, both with a minimum of technical jargon
- Excellent interpersonal relations since your clients are usually resistant to implementing the policies and procedures needed to meet regulations
- Training skills
- Negotiation and conflict management skills
- Common sense and an economic sense are required
- A commitment to environmental protection and safety but not overzealous
- Good organization and prioritizing skills
- A strong ego and a good sense of humor

teach high school. He had never considered teaching, but he decided to accept the job.

When Stockton was offered an opportunity to go to graduate school, he accepted and went full-time, planning to get a PhD and teach college. He was well on the way to his PhD when he decided to get a full-time job. He graduated with an MS in chemistry in 1972, and resumed teaching high school chemistry, while chairman of the science department.

In 1980, Stockton began to wonder how viable his degrees in chemistry would be in industry. He joined Dow Chemical in Freeport, TX as a research chemist in polymers, while satisfying his interest in teaching as an evening chemistry instructor at a local community college.

In 1986, Stockton heard of and eventually took a position as a chemistry and physics teacher and science department chairman at a new high school in Lewisville, TX. While he had planned to work longer in industry before returning to teaching, the opportunity to start up a new program in a new school was too attractive.

Since 1991, Stockton has been the Secondary Mathematics and Science Coordinator for the Lewisville (Texas) Independent School District. In this semi-supervisory role, Stockton develops curriculum, leads instructional improvement activities in the classroom, and administers science and mathematics programs. While he misses the classroom, he took the job because he felt he could have a greater influence on the education of more students in these subjects. One of his goals is to get more application-based teaching in the sciences and math. Application-based teaching relates the subject to the real world to make it more pertinent to the students and improve their problem-solving skills. His industrial research experience influenced his interest in application-based teaching.

Stockton's career goal is to optimize his contributions in his current role. He sees that he is about as far from the students as he wants to be. Advancement administratively would distance him still more.

Career Advice
Requirements for teaching and administrative positions depend on state laws and need to be reviewed before making any career decisions.

Important personal skills and abilities include good communication skills, especially an ability to listen; a strong feel for relevant subject areas and for teaching; and staying current in chemistry and related sciences. Stockton was very positive about his industrial experience contributing to his teaching skills, adding, "You need to be dedicated to teaching, since the respect and pay in industry are not there."

Vice-Provost and Department Chair
Nina M. Roscher

Education
BS in chemistry, University of Delaware; PhD in chemistry, Purdue University

Career Path
Nina M. Roscher was Vice-Provost at American University in the 1980s and is currently Professor and Chair of the Chemistry Department there. Her career path to academic administration and back to teaching and chair in the college and university environment is illuminating.

After Roscher received her PhD in 1964, her husband took a post-doctoral position at the University of Texas. Roscher worked there for two years, responsible for overseeing the freshman chemistry labs and teaching assistants. Her experience in that capacity was pivotal to her career as her first job in academic administration.

After his post-doctoral position, her husband accepted a research position in New Jersey. Roscher's efforts to obtain an industrial job there resulted in some 200 letters of rejection; she characterizes this period as "before affirmative action." In 1968, she was hired full-time as a Senior Staff Chemist at the Coca-Cola® Export Company's lab in New York City.

Meanwhile, Roscher had written Douglass College (the women's college of Rutgers at the time) expressing her interest in teaching. In 1968, Douglass contacted her and subsequently hired her to teach freshman chemistry and quantitative analysis. In 1972, she was offered a position as Assistant Dean responsible for the Registrar, Admissions, and Financial Aid. She also worked with the Associate Dean on budgets. Roscher continued to teach and guide undergraduate research. Roscher felt that she was well mentored by the Dean and Associate Dean during this time.

In 1974, Roscher took an administrative job at American University in Washington, DC, as Director of Academic Administration. Her responsibilities were similar to those at Douglass College, but with a larger scope. As Associate Professor in Chemistry, she also taught one course a year. In 1976, as a result of reorganization, Roscher became Associate Dean for Graduate Affairs in the College of Arts & Sciences.

After another reorganization, Roscher returned to central administration as Vice-Provost for Academic Services. As Vice-Provost, she was the chief academic officer reporting to the Provost. The Provost has the overall daily operating responsibility for the university, similar to a Chief Operating Officer (COO) in industry. The Provost reports to the President, who is responsible for fundraising, representing the university at high level functions, and leading long-range planning. As Vice-Provost, Roscher's responsibilities included academic budgets, the Registrar's Office, Admissions, the Career Center, the Computer Center, the Grants Office and its contracts, the setting of personnel policies, and negotiations with the union of the nonfaculty teaching staff. She took on the added role of Dean of Faculty, which included authorizing teaching contracts, promotion, tenure, leave, and retirement for approximately 500 full-time faculty and an equal number of part-time faculty.

In 1985, Roscher took a sabbatical and afterward returned to the chemistry faculty. In 1991, she became Chair of the Chemistry Department at American University. Her responsibilities include a number of administrative functions, including overall Department planning, developing class schedules and faculty teaching assignments, supervising part-time faculty, overseeing the maintenance of the building and computers, and developing and maintaining the budget. She advises graduate students, deals with office and lab space assignments, chairs the faculty meetings, supervises nonfaculty staff, and assigns graduate teaching assistants.

Roscher is enjoying her current position. However, she would consider another administrative job such as the Provost or President of a small college.

Career Advice
One requirement for academic administration is a PhD. Teaching skills are required since candidates for academic administration usually come from the faculty. In Roscher's case, in several of her administrative assignments she continued to teach a reduced load en route to becoming Vice-Provost. As Chair, she still has some teaching responsibilities.

Personal abilities and characteristics include good interpersonal relations and communication skills. Interpersonal skills are very important since in a university there is a very diverse faculty who have strong opinions or take difficult positions and do not hesitate to express them. "You need to be comfortable with budgets and numbers," she says, "and have a good memory for details and rules, particularly in negotiations."

Exposure to people and experiences outside your department but within the university (and sometimes outside the university) is important if you are interested in academic administration. "Take advantage of opportunities as a faculty member to take on committee assignments," advises Roscher. "You learn what the work is like, whether you like it, earn recognition by a broader group of people in the university and do a good job no matter what." Committee assignments offer opportunities to network, which is important to career advancement.

It is also important to become active off-campus in a professional organization, such as an ACS local section as well as nationally. Roscher has served on, or as chair of, a number of different committees at the national level of the American Chemical Society since 1973. She was President of the Chemical Society of Washington (DC) in 1995.

Provost

J. Ivan Legg

Education
BA in chemistry, Oberlin College; PhD in chemistry, University of Michigan

Career Path

Networking is very important in academic administration and was one factor in the academic administration career of J. Ivan Legg, who is Provost at the University of Memphis, in Memphis, Tennessee.

After a post-doctoral year at the University of Pittsburgh, Legg joined the faculty at Washington State University as Assistant Professor in 1966. Legg did not have administration in mind when he started out in academia, but notes that it was a natural progression to eventually move into academic administration. Even as an assistant professor, Legg developed research programs, obtained funding, attracted undergraduate, graduate, and post-doctoral students, and supervised their work and progress. While the scope was limited, these responsibilities helped him develop leadership and management skills. This leadership role, which he continued for 21 years, and subsequent administrative positions that he held, such as Chair of the Chemistry Department from 1978–86, served to increase his breadth and level of responsibility and experience.

After a couple of years as the Chemistry Department Chair, Legg was nominated for deans' jobs at other research universities. He ignored them at first but over time, he wondered if someone saw something in him that he had not recognized, so he began to apply for some of the positions. He was a finalist in several cases, and turned down one offer in another.

In 1987, he received and accepted an offer as Dean of the College of Sciences and Mathematics at Auburn University. Early in his job at Auburn, he began to receive nominations for the position of Provost or President, but he elected not to follow up with applications. After a couple of years at Auburn, he began to look toward getting a provost position. He had developed an understanding of the system for getting a provost job, began networking in earnest, and submitted applications in response to some nominations.

The nominations resulted from his national exposure in his activities outside the university during his teaching career. Legg was active in publishing in the chemical literature, and gave talks at national meetings, invited symposia, ACS lecture tours, and at other universities. He was also active in committees and boards for the ACS, the National Institutes of Health, and the Council for Chemical Research, an industrial-academic organization. In most of these he became chair of a committee, organization, or board. This not only increased his visibility and enabled him to develop an informal network, but also provided considerable experience in leading and managing groups. The contacts he made were sometimes helpful when he heard of an opening and wanted a nomination.

As a result of his activities, Legg was selected in 1992 for the Provost position at the University of Memphis, a research university with approximately 5,000 graduate and 20,000 undergraduate students. Before accepting, he consulted with four of the research university presidents who had written recommendations for him, and concluded it was a good move and position for him. The move from Auburn after five years was a little sooner than he would have liked, since he felt that he was at a peak and still had some goals to accomplish. However, the opportunity for a senior level position was too attractive to turn down.

Legg's responsibilities as Provost include academic affairs, research, student services, budgeting, and off-campus programs such as televised courses. He has Vice-Provosts who head these various areas and report to him. The split of responsibilities between the president and provost can vary from university to university depending on size, current issues, and the particular interests the President wants to retain. Universities went to the provost structure when they became more accountable for the funding they received from tuition, state and federal governments, and endowments. The result is that the President can spend more time outside the university working and interacting with the public, state and federal legislatures and officials, alumni, and donors, while the Provost is in charge of the day-to-day operations of the university.

Legg is happy in his current job. There are a number of major changes underway at the University that he has a stake in and wants to see happen. He is a member of a group of provosts and vice-presidents of land grant institutions in the US. Their annual retreat, which attracts some 75 participants, is not only a source of ideas and renewal, but also recognizes the level of his career accomplishment and the high quality of his peers at other institutions of which he is a part—a member of a selective inner circle from universities.

Career Advice

One requirement that Legg feels is important for a career in academic administration is to come through the faculty ranks, and become a tenured, full professor. He feels that this enhances empathy and credibility with the faculty. Completing the faculty advancement process is not necessary to hold an associate dean or associate department chair position.

Legg also highly recommends participating in one of the programs that provide official preparation for central academic administration. One example of such a program is the Fellows Program under the auspices of the American Council of Education. Candidates accepted to this program become associates in the office of a university president. They work fulltime for a year in the inner circle of the president and have no responsibilities other than to observe, to write reports, and to share their experiences at conferences with other Fellows in the program. Although Legg did not have such an experience, he feels that his first year in the Provost's job would have been less difficult had he gone through such a program. There are other high-quality programs such as management leadership courses given by some universities. For example, Harvard offers an intensive six-week leadership program.

As for personal characteristics, Legg stated that underlying everything else required to perform academic administrative jobs, it is essential to know how to work with people. You have to be able to listen, and you have to genuinely like interacting with people. The university setting has many talented, tenured, but sometimes very difficult people, and it is important to be able to deal with all of them.

Be prepared to put a lot of "extra" time into the administrative role, Legg warns. Like other upper-management jobs, 60-hour weeks are not uncommon. There are evening and weekend events at a university and a number of social events in which you are expected to participate. In Legg's case, because the University is associated strongly with the city, there are additional outside events, interactions, dinners, and lunches for him to attend.

Important personal assets include the ability to admit you are wrong—which increases credibility. Communication skills are very important, as are conflict management skills.

Environmental Health and Safety

Linda W. Brown

Education
Business administration, the University of Tampa; BS in biology/chemistry, Brenau College; MEd in biology/chemistry, North Georgia College

Career Path
After completing her undergraduate degree, Linda W. Brown gained some background in Occupational Safety and Health Administration (OSHA) regulations while she worked for two real estate development and construction firms over a 15-year period.

Brown decided to change her career direction and become a medical doctor but realized while taking courses part-time at Gainesville College that a career in medicine was out of the question for her; she turned to teaching instead. She was hired as lab coordinator and continued in this role while finishing her BS in biology/chemistry in 1984. Brown began teaching at Gainesville College while completing her MEd in biology/chemistry from North Georgia College in 1987.

Brown has been the Environmental Health and Safety Project Coordinator at Gainesville College in Gainesville, GA since 1993. Brown evolved into the job over a period of time. She was teaching in the Science Department at Gainesville College in 1988 when Georgia instituted an employee's right to chemical specifics, a "hazardous chemical right-to-know" law. Prior to this, most state colleges were exempt from these type of regulations. State agencies were directed to hire a coordinator for this activity and begin training but the law was limited to chemicals at that time. Brown was chosen because she was the junior member of the two chemistry teachers and because while she was a lab coordinator earlier at Gainesville, she had written some lab manuals emphasizing lab safety. Brown had also applied for some grants to do some environmental research on a nearby lake, demonstrating some of the needed technical writing skills and interest in the environment. Since the coordinator's job was one-third time, she continued teaching the balance of the time.

In 1993, the job as coordinator became full-time; her position is considered part of the college administration. Brown's primary task is to assure that the college is in compliance with environmental, health, and safety regulations. Her responsibilities include familiarity with Environmental Protection Agency (EPA), OSHA, Resource Conservation and Recovery Act of 1976 (RCRA), and "Right to Know" regulations, and she conducts right-to-know basic training. The EH&S office employs a part-time person who does office work and publishing. Brown uses the computer to look for changes, or expected changes, in legislation and regulations at the state or federal level in order to meet deadlines. There is ongoing training of full- and part-time and student employees. The training includes not only basic "right to know" education but also education specific to materials an employee is expected to use.

Writing manuals is also part of the job. Safety training ranges from back and lifting safety to ladder safety for the maintenance personnel. As a community college Gainesville College is oriented to serving the community, so Brown publishes a bimonthly environmental newsletter that reaches businesses in the northeast Georgia area. The college also conducts seminars for area industries.

Brown handles calls within the college for any chemical spills and for pickup and disposal of chemical waste including for the chemistry and biology labs, building maintenance for disposal of solvents and paint wastes, and groundskeeping for residual pesticides and herbicides. The art and drama departments also generate paint and solvent wastes, and the photography lab is another source of waste from processing films and prints.

Brown has had opportunities to take other jobs, usually at larger schools or in industry. However, she enjoys her current position and sees opportunities in the area of community outreach. She has been asked by the State Department of Labor to serve on a task force to develop training manuals. Brown foresees that more opportunities will arise in her current position to contribute at the state level and has given a paper on the subject at a national meeting of the American Chemical Society.

Career Advice

Brown holds certificates in Federal Environmental Health and Safety Law and Georgia Environmental Laboratory Safety Principles and Practice, and is a certified Hazardous Chemical Communication Program Manager. While these are not required for her job, she highly recommends taking the courses required to be certified in regulatory law and issues. Certification may be a requirement depending on the job description and laws. Continuing education in this area is a given.

Communication skills are critical; public speaking is part of the job. Brown emphasizes the need for non-technical writing skills, pointing out that it is much different than writing technical papers. Computer skills include word processing and using the Internet to access information. People skills are needed; environmental health and safety work is viewed as a burden on the department or employee, and requires everyone's cooperation and support. Brown recommends taking conflict-resolution training.

Organization and prioritizing skills are needed. These are especially useful in getting funding for the budget, as well as in day-to-day work. Analytical skills are needed for problem-solving. Brown cites the ability to negotiate with people and having common sense, a sense of the economics of the situation. She advises that it is important to be committed, but to avoid being a zealot; it is essential to be optimistic and to be prepared to fight for what you want to accomplish.

For someone entering or currently in college, Brown recommends getting an environmental degree from a four-year school. This route will give certain certifications at graduation in addition to a degree. Certifications in the "OSHA 40 Hour" and "Hazardous Materials Response" are examples. "You need a strong ego so you can stand the job. At times you are not the most popular person when you implement the required changes," she notes.

Brown adds that it is important to enjoy challenge, to have patience, and to be creative in teaching and reaching people. Having a sense of humor can help. Brown's professional activities outside her job demonstrate two other important characteristics: the importance of contributing outside your organization, and the importance of maintaining a network both within and outside of the organization.

— 9 —

Entrepreneurs

An entrepreneur organizes and manages a business undertaking, assuming the risk for the sake of a profit. The happy result for an entrepreneur is a successful enterprise. The word "enterprise" has several meanings which are closely related, many of which describe the characteristics required of an entrepreneur: bold, business, venture, risky, energy, initiative, and active participation. Adding the word "persistence" rounds out the list of personal characteristics and skills needed to be an entrepreneur.

There are many career paths that chemists have taken to become entrepreneurs. Generally they do not start out as entrepreneurs, but have evolved into running their own business after working in industry for some time. There are also chemists who move from an academic or government career to be entrepreneurs. Following are some examples of entrepreneurs and the career paths they have taken.

Entrepreneur

Lyle Phifer

Education
BS in chemistry and math, Wofford College, Spartanburg (SC); MS and PhD in analytical chemistry, Vanderbilt University

Career Path
After receiving his PhD in 1953, Phifer accepted a job at American Viscose in Marcus Hook, PA. The firm had decided to open a pulping plant in Alaska, and most of the analyses of various products for metal ions were done gravemetrically, taking several days. Phifer's first project was to

> ## Skills and Characteristics
>
> - Strong chemistry or other technical background
> - Experience in industry or business and a level of personal maturity
> - Business knowledge and judgment
> - A transition plan that enables a gradual entry and minimizes risk
> - A differentiated quality product or service
> - Customers and potential customers
> - Appropriate facilities, equipment, and personnel to supply product or service at the level needed
> - Problem-solving skills, particularly to get to the heart of the problem quickly
> - Communication skills: listening, oral, and writing
> - Marketing and networking skills, the ability to attract and keep customers
> - Analytical skills, the ability to work with a minimum of information and other times with too much information, much of which is irrelevant
> - Good interpersonal skills
> - Flexibility as the customer's needs may change quickly and the unexpected (good or bad) is common
> - A high level of self-confidence
> - Computer skills at a level required for your particular business or specialty
> - Continuous learning and reading related to your business area
> - Persistence

develop rapid analysis that could give results within 3 hours. Phifer became leader of the Analytical Method Development group at Viscose. Around 1960, American Viscose and Sun Oil formed Avisun, and the Viscose analytical group provided service to them as well.

In 1964, Phifer was approached about taking on a position as part-time chemist working evenings and weekends in a small start-up company. Phifer accepted the part-time job after getting his management's permission with the understanding that his outside job could not interfere with his work at Viscose. The start-up company, Chem Service, packaged very small quantities of a thousand different high-purity chemicals for sale to analytical labs.

In 1976, Phifer was asked to become a full-time employee for Chem Service. After much deliberation, Phifer elected to join the small company full-time, joining the one other full-time employee. The business was successful, and grew about 20% a year. Currently there are 34 employees, with half having at least a BS in Chemistry, two PhDs working full-time, and one PhD working part-time. Phifer is now 50% owner and has full management responsibility.

Today, Phifer is Vice-President and Technical Director for Chem Service, Inc. His main responsibilities are business-related, including insurance, legal problems, payroll, tracking cash flow, employee policies, safety, hiring, firing, employee development, to name a few. "The buck stops here," he says.

Phifer says he may retire in a few years, but he seems to be in no hurry. Two major issues when he retires will be finding his replacement, and how to cash in a portion or all of his share of the business. His grown children are not involved with the business although, in some cases, a younger family member in the business can provide a possible successor. These are typical issues for entrepreneurs as they approach retirement age; some avoid it by continuing to work.

Career Advice

One of the primary factors that led Phifer to become an entrepreneur included networking as a natural outcome of his professional activities. He has also been active in ACS since he started working in 1953 and served on many committees and in several offices, including Chair of the Philadelphia Section of the ACS. It was through the committee activities that Phifer met the entrepreneurs, Jim Jezl and Ed Hollenbach, who originally started Chem Service and asked him to join.

Continuous learning is also important for anyone considering being an entrepreneur. Because of the innumerable chemicals stocked, safe handling of chemicals and safety in the lab is a very important part of the job which requires continuous training and monitoring.

Phifer highly recommends having a good working relationship with a good accountant. He also recommends that in starting up a business, enough money must be in hand to support yourself for several years because you cannot depend on the income from the business to support you in the beginning.

There are no formal requirements for being an entrepreneur, although there are many informal requirements. As an employer, it is essential to know everything: labor laws, tax laws, environmental laws, OSHA regulations, insurance, budgeting, payroll, business, and psychology. He credits his current partner with mentoring him in business matters.

Interpersonal relations and communication are very important in any business; in a small business they are critical. In a small business you can and must treat your employees as persons with families and other interests outside of work. You need to know the employees; understand what, if any, their problems are; and be flexible in handling each situation. Good communication skills contribute to good interpersonal relations. Listening skills cannot be emphasized enough.

Phifer identified the following needs to start and develop a chemistry-related business: a differentiated product or service to sell; customers for the product in a competitive marketplace; sufficient money to support yourself for several years; the assistance of a businessperson; and a good business relationship with a lender, an advertising agency knowledgeable about your market, and a good accountant.

Phifer offers some closing thoughts on starting up a business. "You need a supportive family when you undertake such an enterprise. You can run the risk of sacrificing your family if your family does not understand or agree with your goal. Working 16 hours a day is a necessity and they are not all weekdays. You have to be willing to take on any job in the business and you need to have hands-on skills to do some of these jobs. You can't always afford to hire or wait for somebody to do a repair or a dirty job. You need to do what is necessary."

Entrepreneur

Sallie A. Fisher

Education
BS and MS in chemistry, PhD in inorganic analytical and physical chemistry, University of Wisconsin

Career Path
Sallie E. Fisher's career is featured in this chapter on entrepreneurs although consulting is part of her business. Her business does both analytical and performance testing of ion exchange resins in the lab, as well as consulting.

Fisher's PhD thesis was on ion exchange separation of two metals, rhenium and molybdenum. She taught one year at Mount Holyoke College in South Hadley, MA and the next year at the University of Minnesota-Duluth.

In 1951, Rohm and Haas wrote Fisher asking her to come talk about a job. The company was engaged in ion exchange work on uranium recovery for the US Government. One of the chemists on the project had read her thesis and suggested to management that her experience would be useful. When the government-sponsored project ended in the mid-1950s, she thought about what she was going to do next. Fisher believes that all chemists should take a look ahead every five years and determine what they want to do next.

The result of this planning was that she started building her reputation in the ion exchange field with technical papers and talks. She was tapped to perform some work for the American Society of Testing Materials (ASTM). It was natural that Rohm and Haas, when asked for a representative, selected her since she was developing tests for controlling ion exchange properties before a new resin was manufactured in the plant. Around 1960, Fisher decided that she did not want to work any longer for a large company and took a job at Robinette Research Laboratory, in Berwyn, PA, a small business started by a former Rohm and Haas chemist.

Robinette Lab worked with textile finishes and the textile industry. Fisher was hired to extend and expand the business into ion exchange. Robinette Lab landed a research project in ion exchange for the US Government, work that is still classified. By the time the project was finished in the mid-1960s, Fisher had built up some analytical service business related to ion exchange with other customers, and learned how to manage an independent laboratory.

While still working for Robinette in the late 1960s, Fisher started her own business, Puricons, Inc. In 1972, she went into business full-time and in 1987, Puricons moved to its present location in Malvern, PA. Fisher is President and owner. The name Puricons is a combination of *"puri*fication" and *"cons*ultation". Puricons, Inc. serves users and potential users of ion exchange resins by evaluating, consulting on, and recommending improvements to installed ion exchange systems or designing or recommending new ion exchange systems. Applications include radioactive waste treatment, industrial water purification, and industrial waste water and/or material recovery systems. The work includes certification of materials going into systems, taking samples from plants to re-analyze properties, solving problems with systems, and giving advice and counsel.

Fisher has her own business, but she still considers herself a chemist. This reflects the important technical contribution that she makes to her business. She also handles all the business responsibilities in her present job: she sells, keeps the clients happy, manages staff, prepares quarterly financial reports, outlines research projects, attends meetings and gives papers, and reviews reports going out to clients. Fisher has recently been joined by a second entrepreneurial PhD, and they are supported by a third chemist, a technician, and a secretary.

One creative aspect of Fisher's business relates to employing chemists for short stints along the way. She would hire an unemployed, experienced chemist for a period of time until they got another job.

Fisher's next goal is to carve out the time to write a book on ion exchange. She also has to plan for the future of the business when she decides to retire. It's not clear that she will ever retire; she is having too much fun!

Career Advice

Fisher has had an unusual career in many ways. She started in business at the age of 14, taking over collection of land payments for a real estate business that her grandfather had run. She searched for individuals who had lapsed on their payments and collected the back payments due. She credits learning much about business, communication, and interpersonal relations from this experience that serves her in her present capacity.

Communication, both written and oral, is at the top of Fisher's requirements to run a business. Continuous learning is another key requirement. Financial necessity, the need to have a personal income or to meet loan payments, is a driving force to succeed in business.

Fisher points out that academic training in chemistry today does not create graduates who can, or are motivated to, solve real-world problems, so experience in industry is crucial before going into business as an entrepreneur. For new BS graduates interested in getting a PhD, she feels that working for a year or two in a business before graduate school would be beneficial both during graduate school and in later work.

Entrepreneur

Debra K. Berg

Education
BS in chemistry, University of Delaware, Newark; MBA, Widener University, Chester, PA.

Career Path

In her first job, Debra K. Berg was an analytical chemist with DuPont at the Experimental Station in Wilmington. Her boss noted her ability to start with a lot of data, quickly identify the key pieces, and make decisions, abilities that proved useful in subsequent jobs. Berg started her MBA program while in the lab because she wanted to move toward a business career.

In 1980 she became a Development Chemist on a market development program for elastomer-toughened, fiber-reinforced polyester resin used to mold car parts. She coordinated lab development, manufacturing trials, specifications, customer trials in Detroit, and marketing the product. "I saw firsthand the 'reality' of a new product in the customer's hands and the effort required to get customer acceptance," she commented. Berg maintained a lab responsibility for some of the finished product analytical and physical measurements, particularly when there were product problems.

Berg's MBA work stimulated additional interest in marketing. In what she calls a 'dogleg' in her career, she became a Customer Service Representative in polymer products to get a closer look at sales. Here, she handled the large corporate accounts for a growing plastics parts business, responded to product inquiries, handled orders and customer complaints, expedited orders, and resolved disputed invoices. It was part of an experiment to bring in professional chemists for customer service work to see whether it would be more effective for the business.

In 1984, Berg became a Product Specialist for the nylon filament product line. She prepared sales and earnings forecasts, established pricing, coordinated new product trials, interfaced with and trained international sales representatives, handled customer complaints, and worked with large customers on defining new product needs.

In 1985 Berg and her husband decided to move to California. DuPont offered Berg a position as Senior Sales Representative for pharmaceuticals in San Jose, CA. Berg was not sure she wanted to work in sales, but she had been told for years that she ought to get sales experience as a part of her career. She decided now was the time to do it and learned a lot about technical selling to professionals such as doctors and pharmacists, and dealing directly with customers. She experienced the excitement of a sale, the disappointment of rejection, and the relative freedom of being on her own and planning her day.

In time Berg decided that she did not want to be a career sales person. She left DuPont in 1987 to become a Product Manager with Varian Associates, in Palo Alto, where she was responsible for a new analytical product. She was first charged with developing the product concept, establishing specifications, preparing a business plan, and obtaining top management approval. Her chemistry background, analytical chemistry experience, MBA education, and various marketing and sales experiences, including that of product specialist, had prepared her well.

Berg made the decision to leave Varian in 1989 when she had completed the project. She took a sabbatical and enrolled in short courses on small business and import/export. After a year, she decided to look for a Marketing Manager position and answered an advertisement for a Market Segment Manager at Dionex Corporation in Sunnyvale, CA.

At Dionex, she assumed responsibility for marketing ion chromatography products to the Chemical and Petroleum Market Segment. Her job was to develop market plans, identify customer needs, sell management on the need to develop products for this market, and interact with sales to promote products for this market. Her sales background enhanced her credibility with Dionex sales representatives, and contributed to her success in increasing sales in her segment.

In 1991, because of her interest in needs for on-line chromatography products in the chemical and petroleum plants, Berg became Product Manager for process chromatography. She was successful, but after two years in the job, she no longer felt challenged. She had been a product manager before and was not learning anything new. She had moved away from her technical training. She wanted more business responsibility. She quit Dionex in 1993 and looked at a number of opportunities, but kept seeing product or market management jobs in which she was not interested.

Meanwhile, Berg's spouse had continued his software consulting and development business. Berg began to see some possibilities in his business and they both began to explore how they might expand it. In 1994, Berg became business manager and part owner of The Software Studio, a member of Studio Group 818, Inc., in Cupertino, CA. The Software Studio develops custom software for analytical chemistry instruments and process control instrumentation, among other uses.

As Business Manager of a four-employee company, Berg does the business planning, is the marketing manager, sells, manages the finances, prices products, handles human resource matters, writes proposals, and negotiates contracts.

As an entrepreneur, every day is different. Berg still feels very much challenged by her work. Her current goal is to build the size of the business and revenues so that 25 employees will be needed to handle all the work.

Career Advice
Berg feels that an MBA is a requirement for understanding business. Other requirements for an entrepreneur are past experience in various marketing positions, reading current business publications related to the company's markets and customers, knowledge of the laws related to employment and taxes, and knowledge of contracts and contractual issues. Continuous learning through copious reading and short courses helps her learn and keep up with changes in these areas.

She rated organization skills as the number one personal skill required. In the early years of a business, there are so many balls in the air, one has to be well organized to avoid dropping any. She takes advantage of computer software programs to help keep things organized, develop schedules, prepare Gantt charts, and keep books and finances. Communication and negotiation skills are necessary in her work.

Berg also emphasized the ability is needed to quickly sort through a lot of information, identify key items, and make decisions quickly. Finally, it is essential to be able to handle rejection. Berg says that often when a proposal is not accepted, the potential client does not call to explain why they are not going to accept your proposal—the client simply does not call.

Entrepreneur
Newman H. Giragosian

Education
BS in chemistry, and BS in chemical engineering, Pennsylvania State University; MBA, Wharton School of Business at the University of Pennsylvania; PhD in economics, New York University

Career Path
Newman (Newt) H. Giragosian received his BS in chemistry during World War II and then went into the Navy for two years. When he got out, he decided to return to Penn State on the GI Bill and received a BS in chemical engineering in 1947. His first job was with Allied Chemical in Philadelphia. He wanted to be a plant manager and decided to get his MBA from the Wharton School of Business at the University of Pennsylvania.

Giragosian joined the Shell Chemical Company in New York City in 1951 and was a Senior Technologist for ten years where he did market development and research, and learned that he liked market research and analysis. In 1961, he joined the GAF Corporation and in 1964 became the Director of Marketing Research. His responsibilities included marketing research work, new business and product studies, merger and acquisition studies, research guidance studies, product abandonment recommendations, and various economic studies. While working in New York City, he attended night school and received his PhD in Economics from New York University.

Giragosian had known he wanted to have his own business since he started working after business school. In 1974, he left GAF to form Delphi Marketing Services and today he is the President of Delphi Marketing Services, Inc., and Giragosian Associates, Inc., in New York City. Some of his typical services are product and business development studies, marketing research, competitive and technology assessment, and related services to the chemical industry. He has several technical professionals associated with him and also employs clerical help. Another part of the business is publishing a series of directories, including Custom Chemical Manufacturers, and Custom Compounders. Giragosian's major responsibilities are negotiating contracts for jobs, assigning, coordinating, and keeping current on the work his staff are doing for his business. He also does some of the project work.

Looking forward, Giragosian's goal is to continue to expand his business.

Career Advice

Giragosian feels that industrial or business experience is necessary to go into business or do consulting. That experience sharpens skills such as problem solving and expands knowledge.

The ability to establish rapport is very important in market research where one is asking for information by phone from a person who has never heard of you. Perseverance, analytical skills, and problem-solving skills are needed. It is important to have the ability to conceptualize early in a project what the final result might look like. From this, one can determine the key factors that will impact the results and focus on them. He also cites the importance of having a good memory, so that note taking can be minimized. Additionally, a good memory can be very helpful in making connections to people or information or ideas at the right time.

Giragosian cites networking as being of critical importance for a position in market research for a corporation, and as an entrepreneur, regardless of the product or service. He has served and is active in a number of professional organizations including ACS, the Chemical Marketing Research Association (now Chemical Marketing & Resources Association), Commercial Development Association, and Metro Group of Small Chemical Businesses. In all of these he holds or held various positions such as President, Program Chair, Meeting Chair, Councilor, and Division Chair. He enjoys interacting with people in these various associations and is also affiliated with some eight other professional organizations from the American Institute of Chemical Engineers to the Chemists Club of New York. Giragosian emphasized the vital importance of this networking to him and his business.

He also states that as an entrepreneur, you must be able to work and to enjoy working independently, without the camaraderie of coworkers. He feels that honesty and integrity are essential traits as well. Finally, Giragosian suggests that an entrepreneur should not start out undercapitalized; it is essential to have enough capital to financially survive the first few years as an entrepreneur.

— 10 —

Finance

The area of finance includes understanding where money is coming from and where it is going, establishing budgets, implementing them, and then tracking performance against those guidelines. Generally, the career of a chemist in finance takes advantage of a chemist's technical skills, financial education, training or experience, and interest in working in both areas to meet a need in the employment marketplace.

Many alternative careers list business knowledge as necessary or desirable. Finance is an essential part of business and is an element in making decisions on investing in R&D, newly developed products and processes, manufacturing facilities, and marketing. The direction and work of the chemists and chemical engineers in the lab, semiworks, plants, and marketing are affected significantly by finance and financial considerations. It can be even worthwhile for those who are far from the marketplace to become conversant with financial concepts and terms.

The combination of finance and chemistry or chemical engineering illustrates that even seemingly dissimilar "arrows in your quiver" can work together in an alternative career. It raises the question, "What other combinations of knowledge and experience can I put together to provide an alternative career that I might enjoy and excel at?"

An example of technical skills, finance, and interest combining to fit a niche is the career of Paul K. Raman, a chemicals analyst for a Wall Street investment firm. Raman graduated from the University of Wisconsin with a BS in chemical engineering, and received his MBA at the Manchester Business School in England. He recommends that chemists interested in entering finance obtain a financial position in a chemical firm.

This chapter contains two examples of how some chemists went into financial careers, including camp services supervisor and metals commod-

> **Skills And Characteristics**
>
> - A physical science degree
> - An MBA, or at least additional business training
> - Analytical skills
> - Communication skills
> - Writing skills
> - Interpersonal skills

ity specialist. These may give you a starting point to begin exploring the field of finance.

Camp Services Supervisor
Julie A. Wurden

Education
BS in both chemistry and chemical engineering, University of Washington; MBA in finance and operations management, University of California, Los Angeles

Career Path
Julie A. Wurden's business interest came from her mother, who worked in the real estate business. Wurden became licensed as a real estate salesperson at 18 and received her real estate broker's license at the age of 20. As an undergraduate she worked part-time selling real estate and learned more about business. At the University of Washington, Seattle, WA, Wurden started in the school of business and eventually studied chemistry. She received a BS in both Chemistry and Chemical Engineering in 1982.

Upon graduation, Wurden joined Chevron USA in San Francisco where she had a wide variety of experiences over the next five years. She was project engineer in Marketing Operations, building service stations and replacing underground storage tanks. Subsequently she analyzed economic issues related to transportation and terminal systems for the Chevron-Gulf merger. She then became the Lube Oil Supply Coordinator for moving lube oils from Texas refineries to the northeast packaging and distribution facilities.

As a part of the long-term goals she set when she started college, Wurden took a leave of absence in 1987 to attend the graduate business school at the University of California-Los Angeles. She spent one semester as a business exchange student at Institut Superieur des Affairs in Paris and graduated with an MBA in finance and operations management.

Wurden wanted to round out and formalize what she had learned and apply it to business issues related to her work experience. After receiving her MBA, she accepted an offer from ARCO Alaska (Atlantic Richfield). Since 1989 she has had six different jobs, moving from finance to operations management to use her most recent education. She was an evaluation analyst in planning, operating cost analyst, and a senior financial analyst in the Anchorage office. As a result of expressing a strong interest in operations, she was assigned to work at the production site located on the Kuparuk River 40 miles west of Prudhoe Bay, North Slope, Alaska. First she was the operations support supervisor and then became the materials supervisor, with a broad range of activities and responsibilities in finance and operations with both ARCO and Chevron.

Wurden became Camp Services Supervisor for the 1,000-bed self-contained facility, a job she characterizes as "mayor of a small town." She supervised the catering, housekeeping and janitorial services, security, water and waste treatment facility, airstrip operations, emergency response support, and telecommunications. The typical work cycle was one or two weeks working 12-hour days and then one or two weeks off in Anchorage. There were two occupants in each position, including those employees that Wurden oversaw, who alternated time working and time off. It was a challenge to supervise under this plan since both occupants of a position have to communicate and coordinate to provide continuity.

In 1996, Wurden accepted a transfer back to the "lower 48," moving from operations back into finance.

Career Advice
Wurden states that her MBA has opened the door to opportunities in corporate finance and business. She feels that her technical degree and MBA mean that she is not just a finance person but also has options in petroleum operations, unlike MBAs with a non-technical background. She also noted that a critical factor in getting her current position was that she reminded management that she was interested and qualified as a chemical

engineer to work in operations, a good example of marketing one's experience and capabilities for other jobs.

Wurden felt the MBA was very important for her career path and for learning finance, budgeting, management, and enhanced computer skills. In her supervisory job, she listed the need for communication, organization, prioritization, and motivation skills. She adds that the ability to learn a new subject, like telecommunications, and to learn how to use your education and experience to ask the right questions is very important.

Wurden feels that undergraduate chemists or engineers are not trained very well in business, and thinks it should be mandatory that they take a course in basic finance. She advises pursuing a technical degree, because of the options it offers, but also to take in as much business training as possible, including finance. She felt that the math background of chemists and engineers actually gave them an advantage in learning finance over non-technical business students.

Wurden states that while she has not remained directly involved in chemistry and chemical engineering, she has found a technical education useful because it provides a knowledge base to understand technical issues in oil industry and construction, and in the environmental area.

Metals Commodity Specialist
V. Anthony Cammarota, Jr.

Education
BS and MS in chemistry, Boston College

Career Path
After receiving his MS, V. Anthony Cammarota, Jr. worked for Gulf Research and Development in Harmarville, PA, Ventron (Metal Hydrides) in Beverly, MA, and Sylvania Electric in Salem, MA. In 1963, he saw an advertisement for a chemist at the US Bureau of Mines in College Park, MD. Cammarota interviewed and accepted the offer and became a research chemist at the College Park Research Center of the Bureau of Mines. There, he worked on electrochemistry applied to protective coatings for high temperature materials, new anodes for the electrowinning of aluminum, and processing of metal fractions recovered from solid waste for recycling.

In 1970, Cammarota decided it was time for a change in his career and looked for positions within the Bureau of Mines or the Department of the Interior. Through networking, he learned of an opening for a commodity specialist in the Minerals Information Section. He applied and was hired.

From 1970 to 1982 Cammarota specialized in the commodities of mercury, zinc, cadmium, lead, and tin and acquired increasing technical and supervisory responsibilities. He received training in economics, marketing, mining, and geology, working with the minor commodity mercury and moving up to major commodity responsibilities like copper, lead, zinc, iron, and steel that have greater implications worldwide. While mercury was a minor commodity, he picked it up about the time the mercury poisoning issue arose in Japan.

In 1982 he became Assistant Director for Minerals Information, directing the worldwide minerals information program for the Bureau. Subsequently he held other management positions including Chief of the Division of Mineral Commodities, Chief of the Division of Policy Analysis, and then Senior Technical Advisor. In this last role, he served on professional association committees, participating in quality improvement programs and working on special projects on recycling.

Cammarota analyzed the supply and demand for certain minerals and metals and determined long-range forecasts for the US Government. In his role, he defined factors affecting supply or demand, prepared commodity profiles and long-range forecasts, issued papers to describe important commodity situations, and analyzed the need for Government actions. The specialist is the expert in the commodity area and is a source of information for the Government and industry. The specialist becomes knowledgeable about the sources, countries, political climate, investment climate, mining laws, economics, mining and refining technology and trends, environmental issues, end uses and trends—everything of importance about the commodity during its life cycle.

Cammarota retired from the US Bureau of Mines in 1995.

Career Advice
A physical science degree was needed for Cammarota's minerals-related commodity specialist position, and provides a good base. Different employers in other commodities may have their own degree requirements. Coursework and on-the-job experience met Cammarota's needs in economics, geology, natural resources, statistics, international relations, and legislation. In today's competitive climate, additional education may be needed for one to be hired for a similar job.

Analytical skills are very important by the nature of the work. Communication and interpersonal skills are important to building networks for collecting information inside and outside the Government. Cammarota points out that much of the information a specialist needs is gathered informally at meetings, conferences, and over the phone. Writing skills are needed to report the results of various studies. People skills are also important because some of the work is done by teams or groups with a leader who is responsible for several specialists and also retains hands-on responsibility for part of the work.

Cammarota mentioned that financial institutions like large banks and the World Bank also have commodity and industry specialists on their staff. These specialists keep abreast of their specialty and network with other specialists in the government, private, and academic sectors. When the institution is asked for a loan to open or expand a mining operation, or build a metal refining plant, these specialists are one of a group of experts who are consulted. They provide a report and recommendations based on all the factors from their commodity or industry perspective that could influence the success of the venture and the decision to grant the loan. While there are other factors in the loan decision process, the input from the industry and/or the commodity specialists is a very important factor.

— 11 —

Government Work

There are many positions in state and federal government that can be filled by chemists who wish to use their chemistry backgrounds. State government job postings are often advertised in places like county libraries, federal government job openings can be found on-line and in publications such as *America's Federal Jobs—A Comprehensive Guide to Job Openings in the Federal Government*, and internships and government post-doctoral positions can often lead to permanent positions.

This chapter profiles several positions to illustrate the variety of career options available in the government to chemists. The profiles include positions in law enforcement forensic and crime laboratories, public safety, environmental enforcement, armed forces, county economic development, National Aeronautics and Space Administration, the National Science Foundation, and the State Department.

Law Enforcement: Forensic Laboratory

Daniel W. Vomhof

Education
BA in chemistry, Ausburg College, Minneapolis (MN); MS in analytical chemistry, University of Arizona, Tuscon; MS in manufacturing engineering technology, and MS in occupational safety, National University, San Diego (CA); PhD in biochemistry-plant physiology, University of Arizona, Tuscon

Career Path
Daniel W. Vomhof spent three years at the National Bureau of Standards as an Industrial Research Associate sponsored by the Corn Refiners Association. He joined the US Customs Service to head up the Chicago laboratory

96 CAREERS FOR CHEMISTS

> ## Skills and Characteristics
>
> - A broad background in chemistry, usually with a specialization
> - Computer skills
> - A knowledge of law but not necessarily a law degree; familiarity with laws and regulations and how to interpret them
> - Good communications skills, oral and written
> - A knowledge of how government works at the state or federal level
> - Industrial, or in some cases academic, experience
> - Flexibility
> - Interpersonal skills

which provided technical support and consultation on products, materials, and methods of manufacture to the Regional Director, District and Port Directors, Special Agents in Charge, Special Customs Agents, and importers. While at Customs he worked closely with the Bureau of Alcohol, Tobacco and Firearms laboratory to keep up his analytical skills and learned document handling, examination, and analysis, and detection of narcotics and other controlled substances.

Since he liked the climate in California better than Chicago, he transferred to the San Diego office of the ATF. While in San Diego he read a newspaper article on the use of forensic science at a consultant firm, Expert Witness Services, which was run by a medical forensic scientist. Vomhof contacted the owner and was invited to become a partner specializing in nonmedical aspects of the business. Since he was young and had accomplished his goal of becoming the head of a laboratory, he agreed and eventually bought out his partner.

Much of Vomhof's work concerned accident reconstruction. Since rapid determination of speed from skid marks was becoming important, he bought a computer and learned to program. Vomhof developed software which is used today by attorneys, adjusters, and others involved in litigation, including programs on accident reconstruction, drunk driving evaluation, vehicle databases, paternity blood type/HLA/DNA evaluations, and slip/trip/fall evaluations.

Vomhof was the first expert to use videotape for demonstrative evidence of driver visibility in the courts of San Diego county. He also uses videotape in studies of individual alcohol tolerance and driver performance; these videotapes also have been shown in courts on numerous occasions.

He has provided technical support and consultation to the US Attorney's Office in San Diego, US Customs, San Diego Narcotics Task Force, US Navy and Marine Corps legal officers, the Maine Deputy Attorney General's office, the Santa Ana, CA Police Department and the City of Chula Vista, CA. He has given in-depth training to attorneys on the technical significance of physical evidence, the technical methods used, the technical problems of interpretation of technical data, and the methods of courtroom presentation of evidence. He has conducted seminars for attorneys and police officers on such topics as driving under the influence of alcohol and drugs; vehicle accident reconstruction; controlled drugs; product liability; and homicide and other felony evidence.

Vomhof has prepared over 2,500 technical reports on the evaluation of physical evidence in litigation matters, both criminal and civil, and consulted, evaluated, and advised on an additional 300 cases where no written report was desired. Vomhof has given expert testimony in more than 600 trials and has been deposed in more than 100 cases; in all of these cases he interpreted the technical information and explained the results of his evaluation in the social and behavioral contexts of the courts, the attorneys, and the jurors.

Vomhof also developed and markets an evidence collection kit for collecting and documenting physical evidence found on the victims of sexual assault. The kit is used by hospitals, crime laboratories, and law enforcement agencies throughout the United States. He also teaches courses at National University in San Diego and at Coleman College in La Mesa. Since much of his work is concerned with mechanics and safety, he obtained additional MS degrees in Manufacturing Engineering Technology and Occupational Safety to extend his credibility with attorneys and jurors.

Vomhof expects to continue what he is doing until retirement. Should he decide to leave the business, he would like to continue teaching.

Career Advice

Vomhof says that very little in his career happened by design. He stresses the importance of ongoing education in this era of rapidly advancing science and technology. Vomhof says that beyond a BS degree in chemistry from an ACS-approved department and a broad knowledge of chemistry and physics, knowing how to use computers is essential to be a forensic scientist. He feels that an MS or PhD degree can enhance your credibility.

While good writing skills are important, he also points out that one must be a good oral communicator since many attorneys do not want written reports. It is essential to be able to interpret technical results to the nonscientist.

He feels that it is necessary to have confidence in your abilities but that forensic scientists must be able to subordinate their egos and let others take credit for the results of their work.

Law Enforcement: State Crime Laboratory
Anna L. F. Ezell

Education
BS in chemistry, Mississippi College; MS in biochemistry, Florida State University

Career Path
Anna L. F. Ezell taught college chemistry for almost twenty years before joining the Mississippi State Crime Laboratory as an analytical toxicologist. Her four children were born during the time that she was teaching, and she taught part-time when the demands of motherhood were the greatest.

At the Crime Laboratory, she became supervisor of the Toxicology Section, and was later appointed Associate Director. To supplement her chemistry degrees she has taken courses in medicinal chemistry and computer science, and has received specialized training from the Federal Bureau of Investigation, the Bureau of Alcohol, Tobacco and Firearms, and other forensic organizations.

The crime laboratory employs chemists because much of the work involves analytical chemistry, particularly instrumental analysis. Employees are given on-the-job training in the forensic aspects of their work and in various facets of the criminal justice system. As forensic scientists, they provide training for law enforcement personnel in the services provided by the crime laboratory, proper evidence collection, and crime scene investigation, and may serve as part of a crime scene response team.

Career Advice
Ezell's laboratory work is relevant to her chemistry background because much of the work involves analytical chemistry, particularly analytical instrumentation. She feels that chemistry is an excellent background for

working at a crime laboratory, and states that chemists can be trained in criminology aspects that they will need to know, including how to testify.

Ezell states that an ability to work well with people, and to have good communication skills, are essential. A part of her responsibility is to teach law enforcement individuals forensic science and crime scene response, and how to collect evidence.

Public Safety
David R. Parker

Education
BS and PhD in chemistry, University of California at Davis

Career Path
After David R. Parker received his PhD, he was employed for seven years at a consumer products company, first in new products research and development and ongoing brand support. Here he was introduced to market research, packaging, and process development. Subsequently, he was offered a position in the product safety group where he found new challenges and where he obtained experience in toxicology and regulatory affairs.

Parker decided to make another career switch, relying on his network, ACS Career Placement Services, and scanning the newspapers for advertisements. He saw an advertisement for a chemist with the Santa Clara Fire Department and talked to the two chemists who worked at the Fire Department and liked what he heard, so he applied for and got the job. He felt that it was a small leap from product safety to public safety.

Today, Parker is Administrator of the Hazardous Materials Division of the Santa Clara (CA) Fire Department. He consults with business and government agencies on issues of hazardous materials storage, handling, and use. He reviews building plans for constructing, modifying, or closing facilities using hazardous materials and then inspects the work done on the facilities. He also interacts with the media and balances public safety and right-to-know with corporate and personal privacy.

The three-person division provides hazardous materials emergency response training to the rest of the 150-person fire department; division chemists are members of the emergency response team called to hazardous materials releases. The emergency response function involves using technical knowledge and common sense to abate the hazard, look for patterns, suggest solutions, and refer to the District Attorney for prosecution when appropriate. The division is involved in informing and educating the public and in training firefighters in hazardous materials emergency response, educating them about chemistry, toxicology, and response techniques. Parker is currently working with state and county officials on permit streamlining to eliminate overlap in regulations.

Parker finds what he does interesting and challenging. He feels that he got in on the ground floor and the scope of his job is evolving at a good rate, and at present he is not feeling any urge to move on.

Career Advice
Parker feels that the skills needed for his position are a broad knowledge of, but not necessarily a PhD in, chemistry and toxicology. He has taken university courses in industrial toxicology at Wayne State University and holds a hazardous materials management certificate from the University of California Extension. He has taken ACS continuing education courses on toxicology for chemists, technical writing, time management, and project management. He feels he has gotten a lot of on-the-job training.

Parker sees a need for good written and oral communication skills, especially the ability to translate technical and regulatory ideas and requirements to audiences with little technical knowledge or interest in the subjects. Good interpersonal skills are needed since he works with and within a number of disciplines and with people of varying education levels. He says that a sense of adventure is often useful. At the present time the department prefers chemists who have industrial experience since it can enhance their ability to work with industry.

Environmental Protection
Alan M. Ehrlich

Education
BA, MS, and PhD in chemistry, Massachusetts Institute of Technology; MBA in finance, Georgia State University

Career Path
Ehrlich realized early on that he did not want to work in the laboratory. After he got his PhD and worked at the Coca-Cola® Company, he obtained an MBA in finance. He then entered government service with

the Consumer Product Safety Commission where he worked in regulation development for seven years. Because of his ability to use his technical background in developing regulations and protocols, he moved to EPA to manage a new program for developing policies, administrative procedures, and technical protocols for controlling direct and indirect additives to drinking water.

Ehrlich then moved to the Regulatory Integration Division, Office of Policy Analysis, where he developed strategies for optimal control of pollution from major industries and was responsible for health risk assessment for all of the industries studied by the division as well as for managing studies on organic chemicals and metal finishing. He realized that he liked applying his science background to legal matters and that a law degree would enhance his marketability.

Ehrlich moved to the Office of Health and Environmental Assessment (OHEA) and entered the evening division of the law school at George Washington University. At OHEA he was responsible for program planning and the technical liaison between OHEA and the Office of Solid Waste and Emergency Response. He was the first executive secretary of the Risk Assessment Forum, where he coordinated development of the first round of risk assessment guidelines and was responsible for building consensus on science policy and risk assessment issues.

Upon obtaining his law degree he joined the EPA's Office of General Counsel. He realized early on at EPA that for him, law was fun. He also realized he had a talent for communicating the various legal, technical, and policy issues involved in a regulation, and that he had a talent for drawing people out to determine what it was they wanted or needed from the agency.

While he is not working directly in regulatory affairs today, Ehrlich works in the area of patents as a registered patent attorney, working with agency inventors and outside counsel. He prepares Patent Office submissions, negotiates and/or reviews licensing agreements. He is involved in work related to the Federal Technology Transfer Act, and he negotiates and/or reviews Cooperative Research and Development Agreements. In the area of intellectual property, he provides advice to the EPA. Ehrlich is also an Associate Professorial Lecturer in Law at George Washington University, teaching a seminar in Law, Science and Technology.

He enjoys what he is doing and intends to stay where he is until retirement; then he would like to teach law or go into a patent or environmental law firm.

Career Advice

When asked how he managed to get all the transfers in his career, Ehrlich says that he was the right person, in the right place, at the right time.

Ehrlich feels that the skills or abilities needed to succeed in government are technical sophistication, not necessarily a technical degree; a skill in the law, not necessarily a law degree; a knowledge of how government works, preferably gained from having worked in government; and very good communication skills, especially written communication.

Ehrlich compares the roles of typical scientist and typical lawyer and states that a scientist is used to certainty from experiment and analysis, whereas a lawyer must be able to work with uncertainty and have the self-confidence to make judgments or give opinions while dealing with that uncertainty. While the scientist looks to the past for inspiration, because he or she learns from and build on what has been done, the lawyer looks to the past for justification—that is, precedent.

County Economic Development
Heinz Stucki

Education
BS in chemistry, Ohio State University; PhD in organic chemistry, University of Wisconsin; MBA degree, Rutgers University

Career Path
Heinz Stucki began his industrial career with CIBA-GEIGY in Basel, Switzerland, starting in the patent department and moving into a traditional research laboratory. He came back to the United States and worked for American Cyanamid as a senior process chemist for six years, and then for five years as a business planning analyst, sales representative, and market development specialist. While at American Cyanamid he obtained his MBA degree. He then worked as a commercial development manager and as a technical manger for Hüls America for five years.

Stucki decided to leave corporate life; he and his wife bought a small farm in Ohio and spent a year building a house. He saw a newspaper ad for a half-time position as Director of Economic Development for the Community Improvement Corporation of Tuscarawas County, Ohio. He answered the ad with a proposal for what he would do in that position and got the job. He spends the other half of his time working as a consultant on coatings.

His mission as Director of Economic Development is to advance, encourage, and promote the industrial, economic, commercial, and civic development of Tuscarawas County. Stucki promotes local industry to the media to attract industry to the county. He creates ads, runs advertising campaigns, and publishes a newsletter, which is now on the Internet. The Corporation erects speculative industrial buildings to attract businesses to the county and he oversees the process from construction to the final sale and occupancy by the business. He runs a database on all real estate in the county and is the liaison to regional and state government for economic development. His department tries to attract high-tech companies, focusing on polymers and plastics; Ohio has more plastics companies than any other state. He also chairs the technology committee of the state Development Association. He is also a member of the advisory committee of the Ohio Valley Plastics Partnership whose goal is to attract plastics companies to the Appalachia region of Ohio. He is involved wherever technology and business development come into play.

Stucki is extremely pleased with what he is doing and he looks forward to building on and expanding his current work.

Career Advice
Stucki feels that his combination of technical and business degrees is the key to his success as an economic development director. His chemistry background and polymer experience have been essential to his success because they give him credibility; companies interested in locating in Tuscarawas County are pleasantly surprised to find that he is knowledgeable in their area of technology.

In addition to the technical and business skills, written communication skills are also necessary because he is called upon to write business letters and articles. People skills are also very important since he must work with committees and boards, constantly interacting with diverse groups. Patience to follow through is essential since feedback is often slow in coming.

Stucki's advice to those looking for a career outside their field of technology is to seize every opportunity to gain new experience and new expertise. He feels it is necessary to look for the interface between disciplines because that is where the opportunities lie.

Armed Forces
David C. Stark

Education
BS in chemistry, Rensselaer Polytechnic Institute (NY); MBA, Golden Gate University

104 CAREERS FOR CHEMISTS

Career Path

David C. Stark obtained his BS in chemistry on an ROTC scholarship. When he graduated, the Army was not commissioning officers in the Chemical Corps so he started graduate work in surface chemistry at the University of Wisconsin-Milwaukee. Two years later, the Army was again looking for Chemical Corps officers, and he was activated; after the Chemical Officer's Basic Course he was sent to Europe to activate a new chemical decontamination company.

After a year with the Chemical Company in Europe, Stark became a security and intelligence officer for 16 months and then became a brigade chemical officer conducting training. He finished his tour of duty in Europe and returned to the US to attend the Chemical Officer's Advanced Course. Then he went to the Aberdeen Proving Grounds (MD) for three years to work in the Physical Protection Division in the section concerned with decontamination. While there he used his background in chemistry to work on the army's next generation of protective respirators for which he received a statutory invention registration (the equivalent of a patent) on a design he invented. He was then transferred to Fort Huachuca, AZ to serve as a brigade chemical officer and training officer.

While in Arizona he felt he could enhance his promotability by obtaining an advanced degree. However, after two years in Arizona he was not promoted to Major so he left active duty. He has maintained his reserve status, rising to Lieutenant Colonel and serving his required reserve duty at the Aberdeen Proving Grounds and with the US Army Edgewood Research Development and Engineering Center (ERDEC) working on chemical agent stimulants. Most recently, he has been assigned to work on the Modular Decontamination System.

On leaving active duty, Stark completed an MBA degree. Since he wanted to stay in the chemical defense field, he joined the EAI Corporation, which had just won a mission support contract with ERDEC. He started out ghost writing technical articles for ERDEC personnel. He also does think tank projects in chemical contamination and decontamination. Recently he has been doing work for ERDEC on the chemical weapons treaty and the development of non-destructive evaluation (NDE) systems for chemical identification.

Stark doesn't know where he will decide to go from here. EAI is currently providing management support to a consortium doing environmental restoration on more than 700 former Russian military posts in the former East Germany. The environmental problem involves heavy metal contamination and petroleum-contaminated ground water. EAI is also doing other analysis studies for the army, so Stark's next career move is a matter of determining what niche he wants to carve out for himself.

Career Advice
Stark believes that the skill or trait most needed by an army officer is flexibility; a military officer must be a jack-of-all-trades. Troop officers must be able to motivate indifferent troops, so interpersonal skills are vital. A staff officer must deal with superiors who want answers regardless of the means, so you must have political skills. The army is moving to a mathematical means of decision making, so a background in math is essential; likewise, computer skills are important. Officers who want to advance must have a bachelor's degree; soon a master's degree will be required. He feels that a degree in chemistry is a good degree to have in the army; however, chemical officers rarely make general.

Stark's advice is that the army is not for everyone, but he feels that it can't hurt to give it a shot. Enlisting in today's army is not a lifetime commitment. He feels it is a maturing process and the army is looking for technically trained people. He feels that his service in the army was good preparation for what he is doing now; he probably would not be working for EAI if he had not had experience in the army. Many of the people who work at EAI are military retirees.

National Aeronautics and Space Administration
Jack Kaye

Education
BA in chemistry, Adelphi University; PhD in theoretical physical chemistry, California Institute of Technology

Career Path
Jack Kaye decided that he did not want to spend his career doing quantum mechanical scattering calculations, so he took a two-year position as a National Research Council Cooperative Research Associate at the Naval Research Laboratory in Washington, DC working in atmospheric chemistry modeling. He then went to the National Aeronautics and Space Agency

(NASA) Goddard Space Flight Center as an atmospheric chemistry computer modeler. He is currently manager of the Atmospheric Chemistry Modeling and Analysis Program of the Office of Mission to Planet Earth at NASA headquarters in Washington, DC.

Kaye manages a research program that supports the use of global computer models and analysis of large-scale data sets, especially but not completely satellite data for the composition of the earth's stratosphere and troposphere, mainly for ozone issues. As a program scientist for NASA satellite projects he works on shuttle and NASA flight activities. He is also involved in long-term planning, interagency and international coordination, and some education-related issues.

Kaye has an enjoyable and challenging position. He feels he has an important role representing his discipline and advocating it, so he foresees continuing in his present position. In the long term, he wants to keep his research skills honed so that he does not forget what it is like to do science, which is a concern of scientists in his position.

Career Advice
Kaye feels that a PhD degree is essential to be credible as a principal investigator at NASA. He also feels strongly that a NASA scientist must be comfortable with making presentations, able to distill facts down to a limited number of points and communicate them both in writing and verbally in presentations. Sometimes the material must be boiled down to two or three viewgraphs or a flipchart, and a science program manager must be comfortable with defending a program and asking for resources in a few minutes. He strongly advocates science students taking public speaking classes in college.

For a chemist interested in doing chemical computer modeling such as that done at NASA, Kaye's advice is to develop a strong background in physical chemistry and a broad background in physics, mathematics, and computing, to take more than the minimum number of courses required in physics and mathematics, and to be able to work with computers using the equivalent of scientific FORTRAN plus some basic visualization programs. Some courses or post-doctoral work in engineering, meteorology, atmospheric science/engineering, and/or environmental chemistry also would be very helpful. Kaye also feels that knowing at least one foreign language is helpful, although it is not essential. For Kaye's current responsibilities, Russian is the most useful language.

National Science Foundation

James J. Zwolenik

Education
AB in chemistry, Case Western Reserve University; PhD in physical chemistry, Yale University

Career Path
After receiving his PhD from Yale University, James J. Zwolenik did post-doctoral work at the University of Cambridge, England, in photochemistry. He then obtained a position as a research chemist in the fundamental research group at a petroleum research laboratory. At the same time he was a part-time instructor in physical chemistry in the extension division of the University of California at Berkeley.

Zwolenik decided to broaden his background in order to move more quickly into administration, management, and policy-making. Because he was interested in science policy, he felt that government work might be the best route. His initial contact with the National Science Foundation (NSF) was at an ACS meeting where NSF had a temporary office, so he dropped off his résumé. Subsequently he interviewed in Washington and was offered a position as Associate Program Director for Chemical Dynamics in the Chemistry Section. He managed two research funding programs in chemical dynamics and chemical thermodynamics, totaling $4 million. He eventually rose to Program Director for Chemical Thermodynamics and visited most major university chemistry departments in the US while managing the programs.

Zwolenik then became a staff associate in the Office of Policy Studies, Special Analytical Section, and the Office of Division Director, Division of Science Resources Studies, where he expanded his experience in science and technology policy. He coordinated the competition for and directed the contract on "Science and Technology in the Innovative Process: Some Case Studies"; worked with ACS on the production of "Chemistry and the Economy;" and personally interviewed 40 vice-presidents and directors of research in large industrial organizations on the role of fundamental research in industry, the results of which were published in *NSF Highlights*.

After nine years at NSF, Zwolenik became staff director and executive secretary for the Committee on the Eighth National Science Board Report at the National Science Board (NSB). The NSF reports to the NSB. He recruited, organized, and monitored the progress of the six-member staff that worked with the NSB to determine the policy issues most important to university, industry, government, and nonprofit sectors of the US R&D system. The results were published in a report addressed to the President and the Congress.

He then became executive secretary to the NSB Committees on Planning and Policy, on Policy Formulation and External Communication, on NSF Support of Basic Research in Industry, and on International Science Activities and was chairman of the Coordinating Committee of all NSB executive secretaries.

After five years as special assistant to the NSB, Zwolenik wanted to broaden further into areas that were administrative, managerial, people- and policy-oriented, while still using his knowledge of physical chemistry. He moved up to Senior Staff Associate for Oversight and is now Assistant Inspector General for Oversight, where he heads an office of nine that investigates allegations of scholarly misconduct in science and engineering; conducts inspections at colleges and universities which receive NSF awards; and has oversight responsibilities for the approximately 200 research and education programs at NSF. Many of these activities require interdisciplinary teams including scientists, engineers, lawyers, management analysts, and professional investigators. He is responsible for the oversight portion of the Inspector General's semiannual report to NSB and the Congress, and he serves as the executive secretary to the NSB Committee on Audit and Oversight.

Looking to the future, Zwolenik remains open to opportunity but has no immediate plans to leave NSF. He can see himself working for a private foundation or university where his diverse experience would be valuable.

Career Advice

Zwolenik feels that beyond the technical degree and experience, an NSF employee needs education or experience in management and administration. For example, shortly after coming to NSF he earned a certificate in public administration from the US Department of Agriculture Graduate School and later attended the Federal Executive Institute in Charlottesville, VA. He feels that the scientist in government requires formal knowledge, seminars, or experience in economics and in science policy.

He feels that to work at NSF one needs a deep understanding of how the research and education system for US science and engineering really works, a familiarity with the rules, regulations, policies, and procedures which govern NSF as an independent agency in the executive branch of the US government, and the ability to apply the logical and orderly thinking of science to the complex areas of science policy and administration. Working at NSF requires a broader knowledge of science than working at the bench and the scientist at NSF must be able to interact frequently and effectively with people, and must be conversant with management and administration. While Zwolenik does not rule out working at NSF for recent PhDs, he does feel that experience in research or education and an understanding of how the science and engineering communities work is necessary to be effective at NSF.

State Department
Francis X. Cunningham

Education
BA in chemistry, Brooklyn College (NY); MBA in finance, University of Delaware

Career Path
When Francis X. Cunningham saw the ad in *The Wall Street Journal* for the US Foreign Service, he thought of all the disadvantages: the starting salary would be half of what he was making; the lengthy competitive process of a written exam, an oral exam and interview, and a background evaluation; he was 46 and would compete with much younger colleagues; and the totally different, possibly dangerous, lifestyle. Then he considered the advantages: going overseas, where the government provided housing, medical expenses, and education for his children; learning other languages and cultures; and the adventure of it all. He decided to apply, and as he successfully progressed through each step, he became more convinced that it was what he wanted.

Cunningham had obtained a BA in chemistry and done graduate study in polymer chemistry and in organic mechanisms. He worked as a control chemist, a rubber compounder, an organic chemist with a government laboratory, an assistant product manager for intermediate chemicals, a solid rocket propellant chemist, and then as program manager for high-energy propellant development. While he was at this last job he obtained an MBA in finance. Then, after 15 years in the rocket industry, he started looking for a job, because funding for military and scientific applications for missiles and rockets was cut back. He realized that the chances of finding a position like the one he had were unlikely.

Cunningham was commissioned as a Foreign Service Officer of the State Department, and a vice-consul and third secretary in the Foreign Service. His first overseas assignment was as an administrative and consular officer at the US Embassy in Brussels, Belgium. He issued visas for entry into the US, issued passports, and provided protection and welfare services for US citizens. After two years, he was transferred to the embassy in Manila. Two years later, he received a three-year assignment as an action officer to various United Nations technical programs. In 1981 the State Department loaned him to NASA Headquarters, International Affairs Section, where he spent two years in the international marketing of the space shuttle's cargo bay for launch of foreign telecommunications satellites. He then served as a State Department Intelligence Officer analyzing nuclear proliferation, missile proliferation, and advanced technology issues, followed by a stint as an action officer for US-Spain science and technology cooperation.

Cunningham then served two years at the embassy in Cairo, Egypt, as science and technology (S&T) officer with the diplomatic title of Science Attaché. The S&T officer is expected to report to US colleagues significant technical progress in the countries for which he is responsible. His general responsibilities were to represent the US and its S&T interests in the country to which he was assigned and to represent that country and its interests in the US. The specific functions of an S&T officer depend on the country of assignment. For most countries, the S&T officer is the on-the-spot surrogate for US government agencies, universities, and industries who have or wish to have cooperative technical programs. Depending on the country, the S&T officer may be concerned with the nuclear fuel cycle (France) or irrigation, hydrodynamics and archeology (Egypt). As Science Attaché in Cairo, Cunningham was also required to administer, tactfully and diplomatically, the trilateral (US, Israel, and Egypt) cooperative science program, which is as political as it is technical.

After Egypt, Cunningham served for two years as an action officer on international environmental protection in the areas of ozone depletion and European transboundary air pollution. Then he was assigned to the State Department Board of Examiners, recruiting and testing officer and specialist candidates for the Foreign Service. Although he is now retired, he continues to declassify State Department documents on a part-time basis.

Career Advice

Cunningham states that no formal education is required to be a Foreign Service Officer. Oral and written communication skills are vital, as are analytical ability, leadership potential, interpersonal skills and sensitivity, and intercultural sensitivity. Candidates with an education in history, international relations, or political science do well in the exam. Of course, a science background is necessary for an S&T officer. The ability to speak a foreign language is not essential because the Foreign Service will teach whatever language is needed to do the job. Although an S&T officer is more concerned with policy than practice, an advanced science degree is appropriate for an assignment to world-class science countries like Japan or Germany.

He cites adaptability, absolute dependability, flexibility, and adventurousness as attributes necessary for success in the Foreign Service. He also points out that the Foreign Service Officer must support US policy regardless of personal feelings, and be ready to serve anywhere in the world.

Cunningham highly recommends a Foreign Service career; the salary is competitive; it offers adventure and challenge; colleagues are cultured, informed, and interested in a variety of issues; and it is a good life which commands respect from others and provides a sense of satisfaction in an important job.

Anyone interested in a career in the Foreign Service can write to the Board of Examiners, Department of State, Washington, DC 20520, for a brochure on the Foreign Service as well as an examination application. Scientists working for a federal technical agency may be eligible for a lateral transfer as an S&T Officer, depending on current Foreign Service needs.

— 12 —

Government Relations

Chemists interested in science and technology policy who want to influence change in Congress, state legislatures, and federal and state governments may want to consider a career in government relations. This field requires experience in the ways of government, but this knowledge can be gained from university courses or degrees in science and technology policy. A chemistry degree provides credibility when scientific matters are discussed. In addition, a knowledge of policy and how it is developed, plus excellent communication skills, are musts.

Since chemists in government relations must work with diverse groups of people they must be diplomatic, be able to listen and assimilate what they hear, and must know when not to speak.

The best way to gain entry into government relations positions is through internships and fellowships, many of which are granted and administered by Washington-based associations like the American Association for the Advancement of Science (AAAS), ACS, or the American Institute of Physics (AIP). There are currently 30 Washington-based organizations which offer fellowships. Another path into government relations is chemistry, followed by a law degree, then a fellowship.

Working for a legislator as a volunteer, intern, or fellow provides experience and contacts; the adage about who you know definitely applies in Washington, DC and in state capitals.

Government relations is a job for a select few, but for a chemist with the right background it can be a challenging and rewarding career. Following are two examples of chemists who work in government relations.

Skills and Characteristics

- A background in chemistry or other science
- Additional knowledge or degrees in economics, law, or political science
- Familiarity with how government works and how policy is developed
- A congressional fellowship, internship, or volunteer work can provide contacts and experience
- Listening skills
- Written and oral communication skills
- Diplomacy skills

American Chemical Society
David L. Schutt

Education
AB in chemistry, Calvin College; PhD in physical chemistry, Princeton University

Career Path
While David L. Schutt pursued his PhD in the Chemistry Department at Princeton, he participated in policy debates at the adjacent Woodrow Wilson School of Public Policy and became interested in the subject. Schutt audited economics courses and attended monthly seminars in the physics department on teaching science to the nonscientist. He felt that science policy was important and that scientists could make a contribution to policy. Schutt decided that he wanted to expand his horizons beyond chemistry, and aspired to a position in which he could make decisions for a large business or academic organization.

Schutt applied for and was awarded an ACS Science Policy Fellowship. He studied the impact of NIH funding on chemistry with an emphasis on quantifying the amount of funding, which is not centralized, and in one Institute he helped to establish an advocacy network of about 1000 ACS members who were willing to communicate with members of Congress on behalf of NIH and its funding. He also studied the role of state and local governments in conjunction with the federal government to help to develop research and technology.

After one year of the two-year fellowship, the position of Manager of the Office of Science and Technology Policy (part of the ACS Department of Government Relations and Science Policy) became vacant; he applied for the position and was hired. Schutt is currently working on reports about chemical R&D spending in order to convince Congress that science and technology are important, and that programs that enhance technology are vital. His office is also studying the federal support for R&D in various agencies, supply and demand for chemists, and environmental R&D.

In his present position, Schutt gets to know many people from academia, industry, and government and considers his present job as a springboard to positions in those areas. He is currently pursuing an MBA in order to enhance his knowledge and his marketability, and sees his options ranging from being a financial analyst, to working within a federal agency doing strategic planning or budget analysis, to working in an administrative position in academe.

Career Advice

Beyond a chemistry background, Schutt feels that some knowledge of economics is important, particularly if one is going to study the national budget. A knowledge of political science is also important. He feels it is essential to understand that science is not the most important item on the Congress' agenda.

Written and oral communication skills are of paramount importance, as are diplomacy, an ability to listen to what people are saying, as well as an ability to know what they need to hear and when they need to hear it.

Schutt feels that his job is to bring chemistry to Congress, and Congress to chemists. Much of his activity is focused on interacting with the scientific community to convince them of the importance of getting involved in the public policy process. He gives three to four talks a year at universities concerning the federal budget and its effect on universities. He also encourages members of Congress to visit academic laboratories both to see what research is about and to establish communication channels. He has learned to bridge the cultures of academe, industry, and government.

Schutt advises any chemist with an MS or PhD who is interested in government relations to get an internship or fellowship. For example, AAAS offers 50 fellowships to work on Capitol Hill and there are some fellowships available through the US Agency for International Development (USAID) which are more diplomacy-oriented. He also encourages networking, stating that it is important to know the right person at the right time.

White House Council on Environmental Quality
Kathleen A. McGinty

Education
BS in chemistry, St. Joseph's University, Philadelphia; JD, Science, Law and Technology, Columbia University

Career Path
Kathleen A. McGinty graduated in 1985 with a BS degree in chemistry from St. Joseph's University, Philadelphia. While at St. Joseph's she worked as a laboratory assistant at the Atlantic Richfield Chemical Company (now ARCO Chemical Company) on products designed to suppress coal dust during mining, processing, and shipping, and on water treatment processes. These projects served as her introduction to environmental policies concerning clean air and clean water.

McGinty then decided to study law in order to understand how law and public policy affect the viability of industry. She enrolled in Columbia University's Law School and took many courses in the school's new program in Science, Law and Technology and also continued her study of chemistry and biotechnology in the Biology Department. While at Columbia, she worked at a major law firm specializing in patent, trademark, and copyright law to enhance her understanding of issues concerning commercialization of sci-tech products.

Upon graduation from law school in 1988, she clerked for the Honorable H. Robert Mayer, Court of Appeals for the Federal Circuit, focusing on commercial science and technology since most of the Court's docket involved patent and trademark appeals.

Since her interests lay at the intersection of technology and policy, she applied for and received an ACS Congressional Fellowship. In 1989, she joined then-Senator Albert Gore's staff. At the time, Gore chaired a subcommittee on science, technology and space. As a fellow she worked on science education issues, legislation to enhance opportunities for cooperative industrial research, and initiatives to strengthen patent protection for process inventions. At the conclusion of her fellowship year, Senator Gore asked her to serve as his senior advisor on environmental and energy issues. She served as congressional staff coordinator for the Senate delegation to the United Nations' Conference on Environment and Development held in Rio de Janeiro in June, 1992, as well as an official member of the US delegation to Negotiations on the Framework Convention on Climate Change and the Antarctic Protocol.

At the beginning of his term, President Clinton appointed McGinty to be Deputy Assistant to the President and Director of the White House Office on Environmental Policy. She established the Office on Environmental Policy, providing a mechanism to unite environmental and economic policy. She has now been appointed to head the statutorily created permanent Council on Environmental Quality (CEQ) and is responsible for administering and implementing the National Policy Act of 1969. The mission of CEQ is to objectively analyze environmental policies, and to serve as a venue for open communication among the Administration, the Congress, industry, interest groups, and individual citizens on environmental matters. CEQ has led the Administration's effort to reinvent environmental regulations which are aimed at cutting paperwork, developing partnerships with industry and the states, building trust, and achieving productive harmony and balance among environmental, economic, and social objectives.

McGinty's priorities include resolving disputes, coordinating major federal environmental actions, and helping to reform environmental laws and processes. At the President's request, she chaired the development of several reforms: improving the efficiency and effectiveness of the Superfund program; a plan to protect the ancient forests of the Pacific Northwest while putting people back to work; a restoration program for the Florida Everglades; an agreement for the California Bay and Delta that will protect the environment while helping to ensure the stability of California's water supply; programs to promote environmental technologies and their exports; and common-sense environmental reforms to lower cost while protecting health.

As to her future, McGinty's first priority is to continue to serve the President and Vice-President to the best of her ability. Beyond the White House, she sees a need for people who are well-grounded scientifically and who appreciate the environmental challenge that we face as a society. She may be interested in a sci-tech industrial company involved in issues of environmentally sound uses for technology, or a position in state or local government.

Career Advice
McGinty says that after technical and communications competence, someone working in government relations needs to be familiar with the law,

political science, and the workings and functions of the three branches of government. Interpersonal skills are also essential, and it is necessary to have the patience to listen to and understand the issues people present. She feels that it is critical to the decision-making process that one listen to the concerns of the different interest groups. She also feels that it is important to use one's scientific skills to test assertions against reality; she is continuously bombarded with assertions and must be able to determine which ones to examine to see if they hold up scientifically.

McGinty cautions that for most Capitol Hill jobs there is no regular recruitment or hiring process, and that luck has a lot to do with landing a legislative job. She feels strongly that one of the best ways to secure employment on Capitol Hill is through a scientific society fellowship. She suggests finding a legislator who has expertise in the same types of policy that interest you. For example, Senator Gore was McGinty's choice because of his interest in science and technology, and in research and development.

— 13 —

Human Resources

A scientist working in Human Resources (HR) most frequently recruits scientists or works in university and college relations, but there are generalists who do run the gamut in HR. HR specialists focus on personnel needs including recruitment, outplacement, training, job evaluations, employee policies, employee benefits, and compensation. Some of these functions may be filled not by chemists but by other HR specialists, for example, someone who is certified as an Employee Benefits Specialist.

This chapter includes three examples of how some chemists went into human resource careers, including a career as human resources generalist, in technical recruiting, and university and college relations.

Human Resources Generalist

James E. Brennan

Education
BS degree in biology with a chemistry minor, Holy Cross College, Worcester (MA)

Career Path
James E. Brennan attended medical school for one year before joining DuPont as a chemist in the Explosives Department. After extended training, he was assigned to a plant in Indiana where, over the next five years, he worked as a control laboratory supervisor and as a manufacturing foreman and supervisor.

Brennan was later transferred to another industrial department. During that five-year period, he had a variety of assignments in manufacturing

> **Skills and Characteristics**
>
> - Communications, oral and written
> - Well-developed networks with management, research and plant supervisors, and academic personnel
> - Familiarity with the company's organizational philosophy, policies, research program, and personnel needs
> - Computer literacy for applicant and employee tracking systems, managing data, and responding to employee questions
> - Flexibility and adaptability
> - Attention to detail
> - Organizational and problem-solving skills
> - Leadership skills
> - Analytical ability

supervision, including training supervisor with the responsibility for developing and implementing management training. Assignments followed in other HR capacities, such as recruiting, wage and nonexempt salary compensation, and labor relations and union negotiations.

His next assignment included transfer to a small manufacturing facility in Pennsylvania. The facility was attracting new products and growing rapidly. As employee relations superintendent, Brennan became responsible for all the HR functions and local community relations. During his six-year assignment, the plant grew from 160 to over 500 employees.

Brennan then transferred to a DuPont acquisition in Harrisburg, PA as the employee relations manager. This site employed over 1200 employees and had its own sales, marketing, and engineering functions in addition to manufacturing. Here, with the help of several HR professionals, Brennan's role was to facilitate the integration of this new division into DuPont. The four-year assignment was challenging and exciting, involving open conflict between the two cultures. The integration was ultimately very successful.

Brennan's next position was in Wilmington as employee relations manager reporting to the Vice-President of Electronics. As a resource to four directors and their divisions, he and a staff of four became responsible for management and professional staffing, personnel reviews, compensation, geographic transfers, and other related HR functions. Seven years later,

Brennan became a principal consultant working with project teams to integrate newly acquired companies into DuPont. Functions included re-engineering, downsizing, outplacement, and compensation. Since his retirement, Brennan has been retained by DuPont as a consultant in the same role. Most recently, he has consulted for a British organization, assisting them in restructuring and bringing together several newly acquired laboratories in the US.

Career Advice

Beyond science training, Brennan says further academic training is not essential to work in human resources, but ongoing training and coursework can help to open doors. He believes that manufacturing experience is critical to succeed in human resources and advises getting manufacturing experience for two to three years before making a transition into human resources. Oral and written communications are very important skills in the HR arena.

It is important to ensure that people are treated fairly and are allowed to develop their skills and interests while balancing the needs of the employee with the needs of the corporation. Brennan points out that human resources work is a job with responsibility but little or no authority; you can only persuade, convince, and compromise.

Technical Recruiting

Karen Nordquist

Education

BA in chemistry, North Central College, Naperville (IL); MBA, University of Chicago

Career Path

Karen Nordquist is the manager of technical employment for Nalco Chemical Company at Naperville, IL. Nordquist was first employed by the Nalco corporate research group creating and modifying high molecular-weight polymers for use in a wide variety of industrial applications. She learned polymer chemistry at Nalco through several courses at the Illinois Institute of Technology and during a one-week course at UCLA. The group in which she worked was instrumental in redesigning the entire polymer line from solid powders to a liquid form.

After eight years on the bench, Nordquist took a position in the technical service area of the International Division at corporate headquarters.

During her eight years there, she moved from staff engineer to manager of export sales and service, and earned an MBA from the Executive Program at the University of Chicago.

Nordquist was offered a new position in the International Division Human Resources Department (IDHR) and after a ten-month assignment with the Chicago United Way as Nalco's Loaned Executive, she returned to corporate HR in the technical employment area.

Nordquist's present position allows her to use all of her technical and business background while searching for the next generation of Nalco employees. Her three careers at Nalco have given her the opportunity to learn about a number of different areas of the company and to make satisfying contributions in each area.

Nordquist is looking forward to retirement. She has been a member of the ACS Employment Services Advisory Board Committee on Economic and Professional Affairs and anticipates getting involved in a number of other projects after her retirement.

Career Advice
An HR candidate should have a technical background and a broad knowledge of the company and its personnel needs. A recruiting job requires good communications skills and the ability to work with a wide range of people in all areas of the company. While Nordquist obtained her MBA, she feels that education beyond the bachelor's degree is not necessary to do her job. She suggests that laboratory experience should precede working in HR as a recruiter; she feels that the best path is to begin recruiting while working in the lab, and then transfer full-time to HR to recruit.

University and College Relations

C. Gordon McCarty

Education
BS and MS in chemistry, Wichita State University; PhD in organic chemistry, University of Illinois

Career Path
Gordon McCarty rose through the ranks to Professor of Chemistry at West Virginia University (WVU). While at WVU, he was involved in placing students in industrial positions, including Mobay Corporation (now Bayer Corporation). Because some of his former students spoke highly of his

teaching ability, he began teaching short courses at Mobay. In 1980, Mobay established the position of Analytical Laboratory Coordinator to facilitate technology transfer among the analytical facilities. McCarty was offered the position. As Coordinator he traveled throughout the corporation learning about the company and its analytical laboratory activities.

Due to his academic experience and his knowledge of Mobay, McCarty was asked by the chief technical officer to take on the position of Manager of University and College Relations and Technical Training, reporting to the Vice-President of Administration in the Polymers Division.

The University and College Relations function in industry establishes and develops a liaison between college and university administrators and faculty and the officers and scientists at the business organization represented. The University and College Relations manager is familiar with the research conducted at the educational institution, the candidates being interviewed from each institution, the faculty's technical strengths and interests, the institutional needs of the college or university, and how the business organization's funds and other resources can best be used to advance its recruiting and technological needs. In some organizations, the funds come from Human Resources, research, engineering or business units, while in other organizations the funds come from a foundation set up for administering corporate social investments.

McCarty remains involved in recruiting and has succeeded in "raising the Bayer flag" in universities. He sees his job as "bringing the two parties together." He maintains special university grants from the Bayer Foundation, monitors current relationships and proposes new ones to the Foundation, and monitors endowed professorships and named lectureships. He ensures that the Foundation is spending its money in a manner consistent with the goals of the Research Division and Bayer Corporation in general. He is active in ACS and the Council for Chemical Research as well as serving on two education advisory boards.

McCarty has found it rewarding to combine two careers. While at the present time he has no desire to do anything else, should he decide to leave his current position, he probably would go back to academia where his experience as Manager of University/College Relations would stand him in good stead in furthering relations between an educational institution and industry.

Career Advice

McCarty states that beyond a PhD degree, the requirements for a position in university and college relations are knowledge and understanding of the dynamics in academia, and how colleges and universities work. Academic experience, although not required, is an asset. Considerable experience in recruiting and in interacting with faculty is a must. A knowledge of corporate goals, major research projects, and corporate recruiting is essential. A University/College Relations person usually reports to the Chief Technical Officer and/or head of Human Resources.

McCarty feels that interpersonal skills are essential for the job; he constantly meets with diverse groups of people and must be aware of their sensitivities. He must be familiar with the company's personnel needs and research goals, and the technical strengths of various academic departments. He must be accepted by both the academic staff and the administrators. McCarty feels that the best route to University and College Relations is to move from academia to industry, but an alternate route could be industrial experience combined with a strong link to academia. An industrial person must recognize that in academia, graduate education is the goal, while in industry the goal is getting the job done.

— 14 —

Law

There are several career paths that are available to chemists in law and patent work. Whether a law degree is required depends on the type of work. Scientists who are not attorneys support the intellectual property process in a number of ways, such as patent agents, patent specialists or advisors, patent searchers, patent examiners for the US Patent and Trademark Office (PTO), and private consultants.

Increasingly, patent specialists or advisors (sometimes called patent liaisons) are Registered Patent Agents. Patent specialists who are not Registered Patent Agents do work similar to the agents, but with one very critical and significant exception: they are *not* allowed to practice before the PTO. They may assist a patent agent or patent attorney in preparing patent applications or responses to Patent Office Actions, but the responsibility for the patent application or response is solely that of the agent or attorney. A patent specialist can define or do patent searches, train scientists in patent matters, and work with the scientist and technical and business management to make patent filing decisions.

Patent specialists always have technical training or education. Many times they are chemists who have experience in R&D or another technical function in the organization. They learn about patent work through courses like those given in preparation for the Patent Bar, in personal study, and in continuing education courses. On-the-job training by an attorney, agent, or experienced supervisor, along with experience on the job, increases the patent specialist's capabilities.

Patent specialist jobs are associated with the R&D function in industry and other organizations. However, the number of patent specialists that perform this function without becoming a Registered Patent Agent has been decreasing. Thus, those currently in a patent specialist position but

> ## Skills and Characteristics
>
> - Technical backgrounds like chemistry are required for positions related to patents. A chemistry education combined with a law education enables an attorney to practice more effectively in law related to either technology or the environment.
> - Interpersonal skills
> - Listening skills
> - Analytical ability
> - Questioning skills
> - An ability to see connections from the data provided
> - Excellent writing skills
> - Problem-solving skills
> - Decision-making skills
> - Attention to detail
> - Good judgment
> - Oral communication skills
> - Organization skills
> - Negotiation and conflict management skills

not an agent are either grandfathered because of their long experience and ability to contribute in the role, or are starting out in the position with the intention of becoming a patent agent or a patent attorney, the latter requiring law school.

Registered Patent Agents are technically trained people who have met certain professional requirements, have passed the Patent Bar Exam, and are allowed to practice before the United States Patent and Trademark Office in patent cases. They can file and prosecute patent applications for others in the PTO, although an inventor is allowed to represent himself in filing and prosecuting a patent application. But as in important court cases, self-representation before the PTO is not recommended unless one has had training in the patent law or experience with the PTO. Passing the Patent Bar Exam is also a requirement for becoming a patent attorney and being admitted to practice before the PTO; other requirements for an attorney are discussed in the Patent Attorney profile in this chapter.

Information on becoming a Registered Patent Agent and the dates for taking the Patent Bar Exam are available through the PTO Enrollment & Discipline Section. The mailing address is: Commissioner of Patents and Trademarks, Box OED, Washington, DC 20231. The exam is currently given only once a year.

There are short courses and reference materials for study that can help in preparation for taking the Patent Bar Exam. The Patent Resources Group, 528 E. Main St., PO Box 1249, Charlottesville, VA 22902 (Internet: patentinst@aol.com) offers a six-day patent bar review course. The Institute for Patent Studies, Inc., PO Box 980, 64 Cascade Road, Warwick, NY 10990-0980, advertises patent bar workshops and a bookstore.

Chemists who are trained in law as attorneys may work in any area of the law. Most chemists who become lawyers combine their technical education with their legal training to competitively differentiate themselves from nontechnically trained lawyers and serve market areas needing that combination, such as intellectual property, environmental law, and food and drug matters. Many of these attorneys still enjoy the science in their work, the contacts with technical people, and the continuing science education gained in working on law problems in the technical arena.

Scientists who work in this area can often migrate from one field to another by obtaining additional education, experience, or passing certain exams. For example, it is not unusual for a patent specialist to study and pass the patent bar exam to become a Registered Patent Agent. Many patent attorneys began as examiners in the Patent Office, went to law school in the evenings, graduated, and passed a state bar exam as well as the patent bar en route. The movement is generally toward jobs that have more latitude and higher salary. Although many legal positions are related to patents, this chapter will profile the following careers: one in the US Patent Office, a patent agent in the US government, a patent attorney, and a litigation attorney.

US Patent & Trademark Office

Mary C. Lee

Education
BS in chemistry, State University of New York

Career Path

In Mary C. Lee's senior year of college she saw an advertisement for patent examiner jobs at the US Patent and Trademark Office (PTO). She was hired and was assigned to examine inventions in heterocyclic chemistry. Lee enjoyed the paper research and chemistry involved in the position.

In 1984 she applied for and was assigned to the Office of Automation where she was responsible for identifying chemical search and retrieval requirements for an automated patent system. In this position, each day was different. She worked with contractors with different projects and shifting priorities.

The Patent and Trademark Office has parallel career paths with advancement opportunities in the professional technical examiner role and in management. In 1987, Lee applied for and was appointed Supervisory Patent Examiner. In this position she trained examiners, reviewed and approved the work of examiners, and handled a variety of administrative tasks. These supervisory positions also provide opportunities for special assignments within the PTO or even at the Department of Commerce level, of which the PTO is a part.

In March 1995, Lee applied for and was approved by the Department of Commerce for her current position as Deputy Director for the Biotechnology Group. This group is made up of about 200 examiners and 50 support personnel. Lee shares the responsibility for managing the group with the Director, including resource acquisition, personnel issues, and legal issues such as answering petitions from applicants and their agents or attorneys.

Lee plans to continue to grow in the management role and progress to a Director's position. She enjoys her management role because she can exert a greater influence on the direction and improvement of the organization.

Career Advice

The examiner position requires a technical background specific to the group of applications being handled, such as Lee's chemistry background for chemical-related inventions. A PhD may be helpful but is not necessary.

It is important to have an interest in law. Newly hired examiners receive over 200 hours of training during their first year at an internal school, called the PTO Patent Academy. The Academy also trains all the examiners on issues related to changes in the law and procedures. During office hours, there are also continuing education courses offered to examiners on current patent legal issues, taught by outside attorneys or law professors. In the examination process, considerable prior patent and literature research is required.

An examiner needs to be able to identify an inventive idea quickly, and to quickly make decisions concerning technical and legal issues. While there is increasing teamwork among examiners, there is still a need for an individual to work independently, be a self-starter, and to demonstrate initiative. Other important characteristics include decision-making ability and good communication, problem-solving, and conflict-resolution skills. Proven abilities in leadership, organization, and priority-setting are important in management roles.

Lee advises chemists who may be considering an examiner's job in the PTO to learn more about the job by reading some patents in a chemical area of interest or attending a one-day seminar on patents. If the opportunity arises, visit the PTO search room and conduct a search there. Talk with someone who has been an examiner; a local patent attorney may be able to help.

Lee emphasizes that her work in special assignments was critical in her career. She notes that these opportunities offer experiences in new issues and gave her a global view of the PTO, its mission, and its overall functioning. Lee states that one advantage of the job is that you can leave it behind when you leave work; because of the confidential nature of the patent application, employees are not able to take work home. Lee has been able to strike an agreeable balance between her job and her life outside of work.

Additional information and applications are available from Human Resources at the PTO by writing to Human Resources, US Patent and Trademark Office, Washington, DC 20231.

Patent Agent
Curtis P. Ribando

Education
BS in chemistry, Washington University, St. Louis

Career Path
Curtis P. Ribando had planned to go to graduate school after receiving his BS but decided to first seek a job at the Patent and Trademark Office (PTO) in Washington, DC. He was offered a job with the PTO and felt this would be a good break from academic work; at the time he did not consider patent work as a career.

While at the patent office Ribando took night courses in biology at George Washington University with the goal of attending veterinary school. After four years with the PTO, he automatically qualified for registration as a patent agent without having to take the Patent Bar Exam; this is how he became a patent agent.

As a result of networking, Ribando learned of and applied for an opening in 1976 as a Patent Advisor for USDA in Peoria, IL. Ribando still had an interest in veterinary medicine, and after establishing residence in Illinois, he felt he would have a better chance to get into the University of Illinois veterinary school. He also looked forward to working on ways of promoting patenting rather than turning down applications as an examiner. Ribando got the position and within several months, he gave up the idea of veterinary school and decided to make a career as a patent practitioner.

At the time he joined USDA, his position was a low-profile one. Patenting inventions made by Government scientists was only for defensive purposes, to prevent others from tying up the technology; and there were no exclusive patent licenses granted. With changes both internally in USDA and externally in the law governing inventions made by US government agencies, the position began to grow and become much more meaningful in the USDA. In 1986, Ribando had advanced to Senior Patent Advisor, and in 1993, he became a Supervisory Patent Advisor, responsible for four patent advisors.

Ribando has direct responsibility for USDA patent matters in 16 states. As he did in his Patent Advisor and Senior Patent Advisor roles, he consults with scientists on patenting, prepares and prosecutes patent applications, coordinates non-US filing activities through an outside counsel, and consults with research managers on patenting issues and decisions. As Supervisory Patent Advisor his major responsibility is to supervise four patent advisors who cover labs in 15 other states. Among the five of them they cover patent matters in some 65 research locations.

Ribando likes his current work. His goal is to continue in his current job while continuing to increase his technical and patent knowledge.

Career Advice

A requirement of the Patent Advisory job is to be a Registered Patent Agent, along with an education in science or engineering. Beyond this, a solid background in the English language, both oral and written, is necessary. An agent needs to be creative in defining a invention in writing and in developing convincing arguments based on facts, which may not be obvious without the explanation. This ability is important in making the case for patentability, and for the broadest claims to which the inventor is entitled. Experience in preparing patent applications is helpful when starting out; Ribando learned much of this from his earlier experience in the PTO.

Because of the breadth of the subject matter in which clients of a patent practitioner may work, a chemist in the patent field should have a broad general knowledge of related sciences and engineering, for example, biochemistry, biology, materials science, and chemical engineering. At times, even an understanding of mechanical or electrical engineering is helpful. Specific training in patent practice and subsequent experience is recommended as well as broadening a scientific background through continuous learning and courses. Good interpersonal relations are important since the practitioner works closely with many different scientists and inventors to secure their cooperation and the pertinent information necessary to define or broaden the scope of the invention.

Ribando feels that his experience in the Patent Office was important in his current work in filing and prosecuting patent applications. Having worked on the inside gave him an appreciation for the environment, procedures, thinking, and pressures, and has made his work as a patent agent more effective.

Patent Attorney

David H. Jaffer

Education
BA in chemistry in 1976, Reed College, Portland (OR); PhD in physical chemistry at Cambridge; JD, Stanford University Law School

Career Path
David H. Jaffer was interested in gas phase kinetics by the time he received his BA in Chemistry in 1976, and he was offered a National Science Foundation grant to do graduate study in this area at Cambridge University in England. While his intention was still to study law, the grant was too good an opportunity to turn down. After receiving his PhD in Physical Chemistry at Cambridge in 1979, he entered Stanford University Law School and earned his JD in 1982. He passed the bar exams and was admitted to practice in California and Colorado. He also passed the Patent Bar and was registered as a Patent Attorney permitted to practice before the United States Patent and Trademark Office.

Jaffer took a job with a law firm in Denver that did a lot of environmental work. About the time he arrived in Colorado, the bottom dropped out of the oil boom. Instead of doing environmental work, he did tax and corporate work for three years.

He moved to Silicon Valley in 1985 and began working for Rosenblum Parish & Isaacs, a law firm in San Jose, CA. doing licensing and joint development agreements. These started to die out in the late 1980s due to a downturn in the electronics industry. In the early 1980s, a Court of Appeals for the Federal Circuit (CAFC) was established, the only court of appeals for patent cases from the Federal District Courts. As a result, a larger proportion of Jaffer's legal docket consists of patent filings, infringement opinions, and litigation related to intellectual property. He also does trademark and copyright work. Jaffer is now responsible for supervising five attorneys.

Jaffer's goal is to continue in a private law firm and build up the practice and the size of the group. He is very active in the ACS Division of Chemistry and the Law (CHAL). He is Chair-Elect (Intellectual Property) of CHAL and a member of the 1995 Officers and Board of Directors.

Career Advice
To practice law, it is necessary to pass a bar exam in the state in which you want to work. A degree from a law school is necessary, although theoretically you could take and pass the bar without going to law school. As mentioned before, practice before the Patent Office requires passing the Patent Bar, whether as an attorney or an agent.

Jaffer feels that is important in patent practice to have a business sense of how technology can add value to a company or an organization. This means an understanding of how a business operates, and how a company's proprietary information can create value for the company. Some background or experience in business is useful.

Writing skills are important; technology must be made understandable, whether to a judge who does not have a technical background, or to the patent examiner, who may not have the depth of the inventor in the subject nor the time to try to understand. Clear writing in contracts is necessary; contract disputes arise mainly because the meaning of some part of the contract is unclear. Joint development agreements relating to technology, and especially possible results from the work, take careful understanding of the plans and careful writing to assure a clear understanding of the agreement not only now but also in the future.

Organization, an ability to establish priorities, and meeting legal deadlines are important. In negotiations leading to licensing agreements, listening skills are critical, along with the ability to ask the right questions. It is important to understand what people want and identify what the key issues are both to your clients and to the other party. Every detail cannot be argued if the job is to get done and also if both parties want a good working relationship going forward.

While he did not do this, Jaffer recommends clerking for a judge or working for a corporate law office for six months while in law to learn more about what attorneys really do.

Litigation Attorney

James C. Carver

Education
BS in chemistry, Centenary College, Shreveport, LA; PhD in inorganic chemistry, University of Tennessee; JD, Louisiana State University

Career Path
After two and a half years at the University of Georgia, first in a post-doctoral position, then as instructor, James C. Carver became an Assistant Professor of Chemistry at Texas A&M University for three years.

In 1978, Carver joined the Exxon Research and Development Laboratory in Baton Rouge, LA, as a research chemist. In the early 1980s Exxon was reorganizing every few years. He was in the process of looking for another job when an early retirement opportunity was offered in 1986. After considerable soul-searching, he took the offer and decided to return to an early career interest, law. In graduate school he had worked at Oak Ridge National Laboratories; he had been in academe and in industry and did not enjoy any of these. Carver felt he had run out of traditional jobs for chemists; while he could perform well in those jobs, he did not enjoy them.

Carver started law school at Louisiana State University and graduated with his JD degree in 1989. Through networking contacts, he became aware of an opening in a local law firm when an attorney left to take a law school administrative position. His position in the top 13% of his law class and his technical background were important factors in his being hired for the position.

Today, Carver is a partner in the firm of Taylor, Porter, Brooks & Phillips in Baton Rouge, LA. He practices environmental law doing toxic tort (a wrongful act, injury, or damage, for which a civil action can be brought) and other types of litigation.

He represents companies in contests with the EPA and the Louisiana State Department of Environmental Quality, and with individuals or groups of individuals who claim injury or damage related to some chemical release. Negotiations are a part of the job, since many of the actions are settled before reaching the court. There are more informal than formal negotiations. Carver notes that a good deal of his time is spent trying to convince a client to come into compliance and convince the agency not to fine them for not being in compliance.

Because of his strong chemistry background, Carver is often asked to give opinions on environmental issues when clients are going to acquire a property or an industrial facility. He reviews environmental reports on the property and sometimes visits the site before advising a client. For example, if there has been oil exploration or production on the property, he evaluates the possibility of naturally occurring radioactive materials (NORM) from drilling on the property. He reviews contracts containing environmental or scientific issues and adds or modifies language to protect the client's interests.

Carver interacts with expert witnesses, who are key in many environmental cases. Because of his scientific background, his relationship with expert witnesses may be better than an attorney with no technical background. This background is important when preparing written depositions or for court testimony from expert witnesses as well, because he can better translate the reports for the client and equally important, as a lawyer, can interpret them in terms of the law and the issues at stake.

Carver plans to keep doing what he has been doing because he enjoys it. While there may have been some disadvantages in going to law school later in his career, he felt that his experience in teaching, industry, and life was an important advantage in law school and subsequently in his practice.

Career Advice

In addition to a law degree, Carver states that the ability to write well is key for someone interested in pursuing a law career. Unlike technical writing, which assumes the audience has a certain level of knowledge, in legal writing it is important to assume that the audience has very little knowledge of the subject. An attorney needs to be a good teacher and address the audience at their level of understanding. Liking to read a lot, and an ability to read critically, are also important in the practice of law. The logical thinking processes and problem solving as practiced by chemists are useful in law.

Overall Carver advises that to be an attorney, one should like working with the law. He also states that it is important to enjoy confrontation.

Carver notes that being confrontational does not always have to be verbal, pointing out that there are two kinds of lawyers: There are litigators who handle cases when they are expected to go to trial and there are transactional lawyers, who write contracts, negotiate deals, do tax work, and almost never go to court. Intellectually the latter are confrontational, but their confrontational style is not expressed orally.

— 15 —

Medicine

Most chemists, and college chemistry students, know that pre-medical students take a considerable amount of chemistry in college. During their first three years, pre-med students usually take the same chemistry courses that chemists and chemical engineers take. Doing well in chemistry and the other sciences is important in getting into medical school. Some students who major in chemistry go directly to medical school.

Just as a chemistry degree can be a basis for many alternative careers with a need for a technical background, chemists can be excellently prepared to pursue careers related to medicine that require a science background.

Chemists can work in pharmaceutical sales, medical equipment sales, marketing medical products, and providing technical support for medical equipment. (See the chapters on Technical Sales and Marketing and on Technical Service.) There are also technical service positions to assist, train, and update the users of medical equipment. While specific training is supplied by the employer, most of these positions do not require a chemist to earn a professional degree, such as the MD.

Careers in medicine that require education beyond a chemistry degree include pharmacists, medical technicians in a clinical lab, physicians, and nurses.

Two chemists with careers in medicine were interviewed for this chapter. One is a retired physician whose career shows how the combination of a PhD in chemistry and an MD enabled him to contribute knowledge to the science of medicine. The other is a chemist who got a BS in nursing; her story is very different and demonstrates that one can get a chemistry education and also pursue an alternative career at almost any stage of life.

> ### Skills and Characteristics
>
> - Interpersonal skills, including compassion
> - Organizational skills
> - Attention to detail
>
> *Physician*
>
> - An undergraduate degree in a science like chemistry is helpful to get into medical school, followed by an internship, a residency, and then Board Qualification in the specialty.
>
> *Nurse*
>
> - To be a nurse, an Associate's degree is necessary, as is passing the state board certification exam for the RN license. For advancement, and depending on the career goal, a BSN, MSN, or PhD may be useful.

Physician

Donald A. Roth

Education
BS and PhD in chemistry, University of Wisconsin-Madison; MD, Marquette University Medical-Milwaukee

Career Path
When Donald A. Roth graduated from high school his uncle, a physician in private practice who was also the doctor for a DuPont plant in Cleveland, arranged for him to work during the summer as a technician in a laboratory at the plant. Over the course of five summers, Roth was exposed to and intrigued by the chemists and their work, and decided on a career in chemistry. In 1944, he received his PhD in Chemistry.

Roth enlisted in the Navy and went to Officers' Training School and was assigned to the amphibious forces in the South Pacific for two years. He wrote a book about his experiences as skipper of "LCT 719." After he left the service, Roth visited the Chemistry Department at Wisconsin and was asked to teach freshman chemistry for two years.

Roth's physician uncle in Cleveland did not let him forget about medicine. He enrolled in the Marquette University Medical School (now Medical

College of Wisconsin) on the GI Bill, and received his MD in 1952. After interning one year elsewhere and three years of residency at the Milwaukee County General Hospital, he stayed on to join the Nephrology Department, which deals with kidney diseases. Subsequently, he became Head of the Renal (metabolic and kidney) Disease Section at the Veterans Administration Hospital in Milwaukee, where he thought that he could apply his chemistry. Roth was also Associate Professor and then Professor at the Medical College of Wisconsin during this time.

The combination of his medical and chemistry background was important to moving his career forward. Roth was involved in the early kidney transplants and did some of the first dialysis of kidney patients in Wisconsin, the latter involving chemistry and electrolyte physiology. He has published some 50 papers of clinical and lab work, including the areas of electron microscopy and gas chromatography.

Roth remains interested in publishing; a current interest of his is following up on 40 past kidney donors to determine any long-term effects.

Career Advice
An interest in medical school and medicine is critical in making the decision to pursue medicine as a career. Although clinical medicine took precedence before chemistry in his practice of medicine, Roth has looked at his diversion from chemistry differently than non-chemists, seeking to integrate chemistry into his chosen profession, and this was helpful to his thinking and his work.

As a professor, he was genuinely interested in teaching, he liked students, and wanted to see them do well. He points out that he had to give up considerable personal time and lab work in order to devote himself to teaching. Communication, interpersonal relations, and problem-solving skills are also important in dealing with both students and patients.

Nursing

Dorothy A. Bell

Education
AB in English, Radcliffe College; MA in English, University of California-Riverside; BS in chemistry, Simmons College; PhD in chemistry, Brandeis University; BS in nursing, Simmons College

Career Path
While Dorothy A. Bell received an MA in English, she was still interested in nursing, an early career interest postponed for personal reasons. She con-

sidered studying nursing at Riverside but put that interest aside to raise children. After she and her family moved back to the Boston area, she decided to go back to college to study dietetics and started out at Simmons College in chemistry, and discovered she was good at it. As an undergraduate at Simmons, Bell was also a teaching assistant, and found she enjoyed teaching as well. She received a BS in chemistry in 1978, subsequently earned her MA in 1982, and PhD in 1987 at Brandeis University.

There was a common thread to Bell's life and career interests: She liked teaching and helping other people, particularly women. Her first child was born with Down syndrome, so she was frequently in and out of hospitals, maintaining her contact with the medical profession and nursing. Toward the conclusion of her PhD work, Bell met many nurses and found them savvy and likable. This interaction served to revive her early interest in nursing, but with misgivings, since she was in her late forties at the time. Nevertheless, Bell went to Simmons College to talk with the administrators of the nursing program and was accepted into the program on the spot. At the same time she received unexpected funds that enabled her to get her BS in nursing from Simmons in 1990.

Bell held several clinical nursing jobs; she worked in a hospital, in a home nursing, and most recently as a nurse manager for a private visiting nurse organization. All of these jobs required a license as a Registered Nurse (RN) obtained by passing a state board certification exam.

Currently, Bell is Assistant Professor of Nursing at Lasell College, a women's college in Newton, MA. The requirements for her position include an RN License, a BS in Nursing and an MS in Nursing, toward which she is currently working. Although Bell is new to her current job, she thinks she would like to become chair of a Department of Nursing. Another long-range option that appeals to her is running a hospice.

Career Advice

Skills needed for teaching include communication and interpersonal skills. Bell states that it is important to be able to communicate complex systems accurately to students who are under enormous intellectual and emotional stress. Students worry that they will not pass the Boards to become an RN, or that they will make a mistake and harm a patient.

Facility with numbers, scheduling, systems, and organization are essential; Bell finds that these come easier for her than her peers who have come straight through nursing school.

Bell suggests that prospective nurses and teachers of nursing avoid having contact only with other nurses, because nursing students can get too isolated. She recommends taking courses outside nursing including history, poetry, music, and math.

Bell warns that after graduation, there is a difficult two-year period, similar to an internship, where the work can be fairly menial. Following internship, she says, a nursing path will become clear. She suggests that a nurse with a BS work for several years before starting a master's in nursing (MSN) program, gaining hands-on experience and maturity.

— 16 —
Professional and Trade Associations

Professional and trade associations such as ACS, the Chemical Manufacturers Association, and the Cosmetic, Toiletry, and Fragrance Association hire chemists for a number of positions including work in environmental affairs, regulatory affairs, member and industry relations, state, local, or federal government relations, public policy, and publications areas. Associations seek chemists because they bring the credibility of a science background to the job.

Communication skills are paramount to work for an association since a major part of the job is communicating with members, task forces, committees, Congress, federal agencies, and the public. Analytical skills, to be able to identify and understand issues, are necessary. Flexibility is another important attribute because associations like their employees to move around in the organization.

This chapter profiles chemists who work in the following associations: American Chemical Society, Chemical Manufacturers Association, and Cosmetic, Toiletry, and Fragrance Association.

American Chemical Society

Edward W. Kordoski

Education
BS in chemistry, King's College, Wilkes Barre (PA); PhD in organic chemistry, University of Maryland, College Park; MBA in management, Monmouth College, Long Island Branch (NJ)

Skills and Characteristics

- Advanced degrees are helpful, but may not be mandatory
- Knowledge associated with the work done at the association, such as cosmetic chemistry or regulatory issues
- In addition to a chemistry background, business, computer, technical writing, and analytical skills are helpful
- Oral communication skills
- Excellent writing skills

Career Path

After receiving his PhD, Edward W. Kordoski joined the Dyestuffs & Chemicals Division of the CIBA-GEIGY Corporation as a process development chemist and was promoted to senior process development chemist two years later. While at CIBA-GEIGY, he obtained his MBA. After three years as a senior process development chemist, he spent 20 months on assignment in Switzerland as the technical representative of a multidisciplinary team that developed a user-friendly, interactive computer simulation model of dyestuffs production. He returned to the US to become team leader of the chemical development research group at the St. Gabriel (LA) Textile Products Division of CIBA-GEIGY Corporation for three and a half years followed by nearly a year as quality assurance team leader.

Kordoski decided to try a new challenge in a position where he could combine his chemical industry experience with his business education. He registered with the ACS Career Services Professional Data Bank as part of his job search, and found that the ACS Office of Industry Relations was looking for someone with his qualifications. Kordoski joined the Office of Industry Relations as a senior analyst in early 1993. He was responsible for expanding the existing ACS portfolio of programs, products, and services for industrial chemical scientists by creating new programs, publications, and conferences. He also worked closely with ACS divisions to enhance the image and stature of industrial chemical professionals.

Kordoski recently joined ACS administration as Manager of Library Services, responsible for determining how the library can better serve its customers, and overseeing the process of becoming a virtual electronic library by using emerging technologies.

Kordoski feels he should be able to advance in ACS administration; he states that his degrees coupled with his experience in the chemical industry, pursuit of continuing education courses, and diverse skills make success attainable.

Career Advice
Kordoski believes that to be effective at a professional association like ACS, the scientist must have, in addition to a scientific degree, an understanding of the English language and grammar and their effective use in writing, knowing how to give speeches and presentations to a broad spectrum of audiences, logic and analytical thinking, basic business skills in marketing, promotion, accounting, and budgeting, effective computer skills, and good interpersonal skills. While an MBA degree can provide a decided edge, Kordoski feels that many of these business skills can be obtained through correspondence courses, short courses, reading, audio and videotapes, televised learning, and other avenues.

He would advise someone who is contemplating working for a professional organization to get a good background in chemistry and to invest in the future by getting a good liberal arts education, including business courses. He sees the key to any job as a good basic science education coupled with diverse, flexible, and portable skills needed in the job market.

Chemical Manufacturers Association
Carolyn Leep

Education
BS in chemistry, Valparaiso College (IN); MS in organic chemistry, Stanford University

Career Path
While Carolyn Leep was working toward her PhD degree in chemistry, she decided that she did not want to work in a laboratory position. Instead, she completed her MS in organic chemistry and took an internship with a law firm and was introduced to environmental and legal issues.

Leep moved to Washington, DC where she took a job with an environmental consulting firm, and became involved in risk issues. She decided not to pursue a law degree and in 1992 she joined the Chemical Manufacturers Association (CMA) as Associate Director of Risk Issues in the regulatory affairs department. She works with groups of member company representatives to develop positions on human health risk issues of member company products and operations. CMA works to advocate those positions with regulatory agencies, and the regulatory affairs department works with the federal and state government relations department of CMA to advocate those positions with Congress and state legislative bodies. They also work with member companies to communicate and educate on the issues under study.

Leep very much enjoys what she is doing. She enjoys the breadth and diversity of contacts she has, both in industry and in government. In the future, she sees herself moving up into management within CMA or another association, or going to a job related to her area of specialization in industry or in a consulting firm.

Career Advice
Leep feels that a technical degree is important to work for an industry association because it prevents one from being intimidated by the issues. While an advanced degree is not mandatory, it can increase credibility and does help toward professional advancement.

She feels that communication skills are vital because one must be able to write or speak clearly to advocate CMA positions, and to be able to work with diverse groups of people to reach consensus on positions. Analytical skills are essential in regulatory affairs to be able to sort out and understand complicated technical and policy issues to arrive at positions. She has acquired a broad-based knowledge of regulations, particularly those related to human health risk of member company products. She feels that networking is vitally important to stay informed on issues.

Cosmetic, Toiletry, and Fragrance Association
Joyce F. Graf

Education
BS in chemistry, University of Vermont, Burlington; MS in organic chemistry and PhD in organic photochemistry, University of Massachusetts at Amherst

Career Path
After obtaining her MS in organic chemistry, Joyce F. Graf became a quality assurance coordinator in a company manufacturing drugs, cosmetics, and medical devices. She completed a PhD in organic photochemistry, using gas chromatography and other instrumentation. Due to family commitments, she held temporary teaching positions, both in secondary school

science and as a college chemistry instructor. She worked at the National Institute of Standards & Technology as a Guest Scientist in spectroscopy and organic synthesis for three years.

Graf realized that although she liked many aspects of research and teaching, she wanted a new career direction. Graf developed a list of activities she enjoyed, skills she had acquired, and achievements. She applied for an opening for a staff chemist at the Cosmetic, Toiletry, and Fragrance Association (CTFA), and launched a new career direction. She is currently the Director of Environmental Science for CTFA.

At CTFA, Graf provides administrative and technical support to CTFA staff, industry task forces, and three standing committees of CTFA; directs programs in environmental and occupational safety and health, air quality, and microbiology, sanitation, and product preservation; monitors information, scientific literature, and regulations; communicates program status at meetings and in technical and association newsletters; and developed and manages the CTFA Hazard Communication Program.

When she is ready to retire from CTFA, Graf would again like to redirect her interests. She will go back to her self-analysis and skills list, which she feels everyone should update on a regular basis. Graf notes that many individuals who come out of trade associations go to state agencies, the FDA, or into consulting.

Career Advice
A career in a trade association requires a good technical background because it increases credibility. Courses in technical writing or journalism, business or marketing, and computers are helpful. Graf also thinks that experience or training in toxicology, quality control, quality assurance, occupational safety and health, and environmental management can be helpful in her position. Familiarity with cosmetic chemistry would be a decided advantage for dealing with cosmetic industry scientists.

She feels that a direct route to a trade association would be via law school or internship work at an association. Other ways to get into a trade association are to work for government, particularly in the regulatory area, or for a company within the industry.

Graf feels that all her life experiences have been helpful in securing her current job and succeeding at it. Having worked with various state and community organizations including the American Association of University Women (AAUW), Parent-Teacher Associations (PTA), and Boy Scouts gave her experience on working on committees and in consensus-building with diverse groups.

— 17 —

Quality Assurance and Control

Traditionally, there are three tasks which are integral to the quality function. The first task is acceptance: to approve a good product based on a set of standards and set aside out-of-specifications product for rework or disposal. The second task is prevention: to statistically analyze the data generated by inspection to prevent out-of-specification product by remedial adjustments to the process. The third task is assurance: to use customer feedback and quality audits to indicate areas requiring remedial action or process adjustment to prevent quality from varying beyond established standards.

These three functions may be assigned to one department or to three separate departments; for example, acceptance to a control laboratory, prevention to a quality control individual or group, and assurance to a person or group that may be responsible for Good Management Practices and for ISO 9000.

The three quality functions reviewed in this chapter are quality control, process improvement, and product quality, integrity, and safety. These profiles should give you a starting point to begin to explore the possibility of careers in quality assurance and control.

Quality Control

Ron R. Richardson

Education
BS in chemistry, Eastern Montana College, Billings

150 CAREERS FOR CHEMISTS

> ### Skills and Characteristics
>
> - Training and experience in technical fields like water analysis, atomic absorption, mass spectroscopy
> - A working knowledge of computers and statistics
> - Good communications skills
> - Familiarity with Occupational Safety and Health Administration (OSHA), Resources Conservation and Recovery Act (RCRA), and Clean Air Act regulations
> - Ability to work with multidisciplinary teams
> - Attention to detail

Career Path

After Ron Richardson obtained his BS, he went to work as a bench chemist in an analytical laboratory in Wyoming. After a year, he moved with the laboratory to New Mexico. Two years later he had the opportunity to become a lab technician at Barrick Gold Corporation Goldstrike Mine in Nevada and took it; two years later he was promoted to chemist.

As time passed, Richardson found himself doing more quality control (QC) work, including setting up the tests, and analyzing and interpreting the results. Management decided that it needed a QC supervisor, so he was appointed Quality Control Supervisor. Richardson sets up and monitors the tests, does the statistical analysis, and then reports the results.

According to Richardson, quality assurance is a system or the programs you put in place, while quality control is used to monitor the programs. Quality control is the testing while quality assurance is the statistical analysis of the quality control results to make sure that your product is within specifications. Quality control is an active process; it assures the quality of the product.

Richardson has recently accepted the position of Chief Chemist at Golden Sunlight Mine, Placer Dome, Inc., in Whitehall, MT. He is excited and eagerly anticipating the challenges this position will bring.

Career Advice

Beyond a chemistry degree, Richardson feels that the only other coursework needed is in statistics because they are used so heavily. Computer skills, mathematics skills, and writing skills are also necessary. He states that someone doing his job must be well-rounded and versatile.

Process Improvement

Timothy L. Guinn

Education
AA in chemistry, Northeast State Technical Community College

Career Path
Timothy L. Guinn started his career in the laboratories of the Organic Chemicals Division of Eastman Chemical Company at the bench where he learned statistical design, experimental control, and process improvement. When the opportunity arose he moved to the technical staff of the plant. His current title is Technician Associate, where he trains the plant operators in process control.

Process control has its own computer program that records data from the plant, reagents, batch weights, times and temperatures when changes are made, and analytical data. These data can then be transferred to a system where correlations are made; for instance, examining the effect of addition rates on the color of the product. Commercial software is also used to do statistical analyses of the data. The operators are involved as much as possible in data gathering and evaluation, and in making decisions based on the data analysis. As soon as a batch is out of control statistically, the operator can begin to investigate why, as part of a continual improvement process.

Much of what Guinn does is examine what is expected from the process and from the operator. He writes procedures and keeps up with government regulations, safety rules, and equipment. He also teaches data interpretation to the operators.

In the future, Guinn sees himself going back to the laboratory or into plant management. The plant experience he has now is invaluable. The next career step is to building supervisor, which involves managing a number of processes.

Career Advice
Guinn's advice is to get as much education as possible in chemistry, computers, and statistics. He sees education as getting a foot in the door; it is

then necessary to keep studying in order to master computer science or management skills.

Most of the statistics he has learned in-house, although the company does offer outside courses in-house. Some of these courses are taught by statisticians and some are taught by computer people, particularly those involved with the programs they develop. Guinn has taken a few formal statistics courses outside the company. Computer skills, technical writing skills, and communication skills are vital to his job since he must be able to communicate with the plant operators and with other technical people. He must also be familiar with OSHA, RCRA, and Clean Air Act regulations. There are employees within the company who interpret these regulations to ensure that the company is in compliance.

Product Quality, Integrity, and Safety
Robert A. Kingsbury

Education
BA in chemistry, Blackburn College (IL)

Career Path
After Robert A. Kingsbury obtained his BA in chemistry, he did a year of advanced study in analytical chemistry. He then went to work for Underwriters Laboratories, Inc. When most people think of Underwriters Laboratories, Inc., or UL, they think of the UL label on electrical equipment meaning that the item is safe to use under normal conditions, and has been tested by electrical engineers. But UL means more than electrical testing. Kingsbury is now a senior staff chemist and manages a group responsible for various tasks including small-scale flammability testing of solids (mostly plastics and fabrics), liquids and gases; identifying materials using infrared spectroscopy and thermal analysis so that the item made out of the material (such as a TV or microwave) can get or maintain its UL label; testing thermal conductivity of low-density thermal insulation (mostly cellulosic) to guarantee the quality of the insulation in the walls of a house; and evaluating flammability of new, alternative refrigerants.

Much of Kingsbury's recent work has been on alternative refrigerants, which has become important recently because of ozone depletion. UL is currently expanding because of environmental concerns and they are now doing work in water quality and sanitation.

Kingsbury enjoys his job and expects to stay in his present position or move up the management ladder at UL.

Career Advice

A chemist wanting to work at a laboratory like UL should have training and experience in technical fields like water analysis, atomic absorption, mass spectroscopy, and supervisory experience in order to advance in management.

Beyond a chemistry degree, Kingsbury states that verbal and written skills are valuable. Oral communications skills are necessary; he conducts many of the guest tours at UL and says he must be able to translate technical terms and tests to the layperson. Since he represents UL on eight technical committees, he must be able to work effectively with multidisciplinary teams. Kingsbury also feels that attention to detail is very important and he traces that skill to his chemistry coursework, which taught him to be careful and precise in his work.

— 18 —

Regulatory Work

Regulatory chemists work in the areas of development, manufacturing, academia, and government ensuring that regulatory standards are met. In carrying out their duties, regulatory chemists perform a number of tasks, including assessing, communicating, and managing risks and liabilities; designing appropriate control and prevention policies; ensuring compliance with and monitoring the status of local, state, and federal laws and regulations; and developing programs and training to ensure compliance with laws and regulations.

There are college programs, short courses, and training programs in various areas of regulation and regulatory law. The chemists interviewed in this chapter learned how to be regulatory chemists on the job. The areas of regulatory work profiled here include pharmaceutical development, manufacturing, academic environmental health and safety, environmental service, clean air (state government), and clean water (corporate business).

Pharmaceutical Development

Bonnie Charpentier

Education
PhD in biochemistry, University of Houston

Career Path
Bonnie Charpentier worked at Procter & Gamble (P&G) for nine years as an analytical chemist. While at P&G she worked on some compounds that had raised some interesting regulatory questions. It became apparent to her that it was important to have scientists involved in regulatory work, so she

Summary of Skills And Characteristics

- A background in chemistry is essential; additional courses such as public health medicine, statistics, political science, environmental law, geology, and biology are helpful
- A working knowledge of the laws and regulations applying to a specific area of responsibility and ability to analyze and reason through highly complex laws and regulations.
- Must respond to unusual situations and schedules
- Ability to work with diverse groups of people and develop training programs for adult learners
- Must be able to interpret laws and regulations for the layperson
- Professional certification (i.e., Certified Hazardous Materials Manager, CHMM)
- Knowledge of computers and computer databases is essential

started to learn about the field at P&G. About the same time, her husband took a position in biotech patent law, and they moved to California. Through ACS contacts she learned about available positions at Syntex and joined that firm as manager of regulatory affairs in the chemical manufacturing and controls group. She worked mainly on chemical aspects of drug development and subsequently her responsibilities expanded to include work on all aspects of preclinical and clinical regulatory affairs, advancing to the positions of senior manager, program director, and then director. She is involved with potential new drugs from their inception, which involves teamwork and considerable paperwork.

Syntex was later acquired by Roche; Charpentier is now director of regulatory affairs for Roche Global Development-Palo Alto. She heads a regulatory team that provides support for drugs currently in development in Palo Alto. She has world-wide responsibility for regulatory approval of the drugs she works on. Her team works closely with the FDA and with several teams outside the US to obtain regulatory approvals for the use of the drugs around the world.

Charpentier loves regulatory work; she sees it from the viewpoint of the scientist, not of the bureaucrat. She enjoys combining science with the strategy of drug approval. She hopes to continue regulatory work in the foreseeable future and especially finds it rewarding to see a new drug make a difference in the lives of patients who benefit from it.

Career Advice
Since there is almost no training at the university level in regulatory work, most people in the field have started out in the laboratory or clinic and moved into the regulatory area. Charpentier says that there is legal training for attorneys who do regulatory work, but training for regulatory affairs is primarily on-the-job. There are short courses on specific issues like basic food and drug regulations or how to prepare and submit an Investigational New Drug (IND) application. What is needed is a solid grounding in a chemical or biological science (chemistry, biochemistry, toxicology, or medicine, for instance). The scientist must be able to communicate with chemists, toxicologists, clinicians, regulatory authorities, synthetic chemists, and formulators, and understand enough to recognize what the issues are in these various fields.

Charpentier feels that in order to work in the regulatory area, one must be able to handle stressful situations, analyze situations rationally and scientifically, and work with a variety of people from different disciplines to form effective teams. The regulatory professional must be flexible, be able to work with details, as well as be able to step back and make sure that the overall strategy of getting a drug approved hangs together at any period of development.

Manufacturing

Carol Jean Bruner

Education
BS in chemistry and education, Beaver College, Glenside (PA); AB in business, The Community College of Philadelphia

Career Path
After Carol Jean Bruner received a BS in chemistry and education with a major in biochemistry, she worked at a medical research institute as a research technician for two years, followed by a position as research assistant at a medical school. While she was there, Bruner took graduate courses in radiation safety.

Bruner took time off to start a family. When she was ready to return to full-time employment, she found a position at a small specialty chemical company by networking with her contacts on an ACS local section committee. She was responsible for setting up the testing program for an FDA-regulated material and in less than two years she was running the quality control department. While Bruner worked in quality control, she went to night school and got an associate's degree in business from a local community college.

After several years, Bruner was also responsible for teaching courses to employees on safety and right-to-know issues. She took on more duties connected with regulations and became the regulatory affairs manager. After 15 years with the company, Bruner is now Vice-President of Health, Safety, and Environment.

In her position as head of regulatory affairs, Bruner keeps abreast of government regulations as they apply to what the company is doing, including everything that requires permits, particularly air and water. She must be current with FDA, OSHA, and workers' compensation regulations. She chairs the safety committee, is responsible for creating material safety data sheets (MSDS), state-mandated right-to-know labels, and is the training director. Bruner also handles hazardous waste disposal. As with many small companies, she wears numerous hats and rarely wears the same hat for more than ten minutes at a time, which she feels makes the job interesting.

Bruner was just promoted to Vice-President of the company and she has no intention of changing jobs in the near future. In the long run, she looks forward to training her successor and then retiring.

Career Advice
Bruner advises anyone contemplating working for a small company in regulatory affairs to have a sound analytical background. While regulatory affairs training should be on the job, she expects to train a new person both on-site and off-site in safety issues, including hazardous waste handling and disposal. She states that being a certified industrial hygienist or certified hazardous materials manager, achieved through coursework and an examination, would be a real plus for someone aspiring to work in environmental or regulatory affairs.

Bruner's coursework in business included computer programming, which she still feels is most valuable. As for other skills required in her position, communication skills are also important because she is constantly writing reports and policies. She values her people skills especially when conducting safety and training since she must be able to handle production people, most of whom have only an eighth-grade education. She notes that her company is very supportive of her working with ACS and local education organizations since they are good public relations for the company but also because they enable her to network effectively on a technical level with her scientific and business peers.

Academic Environmental Health & Safety Manager

John S. DeLaHunt

Education
BA in chemistry, Colorado College, Colorado Springs

Career Path
John S. DeLaHunt wanted to stay in Colorado Springs following his graduation, and since the chemistry department was moving into a new structure, he was hired to help move equipment and chemicals into the new building. To obtain a building occupancy permit from the fire department, DeLaHunt found it necessary to rearrange the chemicals in the storeroom by hazard class. Additionally, a large number of unused and unusable chemicals had accumulated in the chemistry department and had to be disposed of responsibly. DeLaHunt was hired to develop a disposal plan and then to do the actual disposal work.

That job evolved into a full-time position of chemical safety coordinator. He was responsible for collecting and disposing of, or arranging for disposal of, chemical waste from all campus departments; sorting chemical inventories to comply with Universal Fire Codes; serving as the chemical release responder; and handling other chemical safety issues, such as acid storage in the laboratories.

Two years later, DeLaHunt was further responsible for hazard communication, radiation safety, and chemical hygiene according to OSHA's laboratory standards. Three years later he had taken on additional environmental and safety responsibilities, achieving the title of Environmental Health & Safety Manager for Colorado College. DeLaHunt created his own position and was a trendsetter; today many small and medium-sized schools have begun hiring environmental and safety staff.

He collects and arranges for the disposal of chemical and radioactive waste from all campus departments; DeLaHunt still does much of the disposal work himself along with student help, spending about 20 hours per week on waste disposal. He wrote a computer program to track and handle waste disposal from request through final disposal. He also manages radioactive materials as Radiation Safety Officer; serves as Chemical Hygiene Officer according to OSHA's laboratory standards; serves as chemical, radioactive, and biological incidental release responder; assists in resolving ongoing indoor air quality problems, including managing asbestos and PCB removal contracts; chairs the physical plant safety committee; and arranges loss control and safety training.

Someday DeLaHunt would like to get his career writing fiction off the ground, but until that happens he expects to continue trying to make the future of Colorado College his priority by providing protection from curtailment by environmental health and safety concerns or problems. He likes the challenge of what he is doing and enjoys the support of college administration because he is moderate and flexible.

Career Advice
Beyond his BA degree, DeLaHunt has taken classes on radiation protection, OSHA regulations, compliance and emergency response, and is currently taking a course on indoor air quality. He took a chemistry course at the college on transition metals since much of his disposal work involves transition metal compounds. DeLaHunt finds that he needs a broad knowledge of chemistry, and finds he does significant library research because many disposal jobs are not routine. Critical thinking is an essential skill, and so is an ability to see the big picture.

DeLaHunt thinks there will always be a need for environmental health and safety management in colleges and universities. His advice to anyone wanting to follow in his footsteps is to find a college without an environmental health and safety person and create the position.

DeLaHunt foresees his area becoming an administrative department of risk management that would encompass not just environmental health and safety management, but would also include insurance and worker's compensation administration.

Environmental Services

Curt Clowe

Education
BA in chemistry, Hastings College, Hastings (NB)

Career Path

Curt Clowe's first job as a chemist was at the bench doing research. He then became a production-level analytical chemist, followed by a job conducting quality assurance and quality control testing. While working as a quality assurance manager he started doing EPA and OSHA compliance work for a pesticide formulator. He moved on to a position as a corporate-level environmental and safety specialist. He currently works for one of the world's top environmental engineering firms as In-Plant Services Manager.

Clowe manages a group of nine people who do field work, laboratory contracting, data review, regulatory compliance, health and safety, and a wide variety of in-plant services. As an in-plant services manager, Clowe is responsible for providing client-focused services involving environmental management, including auditing, policies and procedures, risk identification and management; hazardous materials and wastes, including minimization, permits, management, regulatory compliance; air pollution, including Title V audits and certification, modeling, and regulatory support and permits; water pollution, including prevention, permits, waste water treatment, and spill planning; and plant engineering, including process safety, hazard analysis, feasibility studies, process and plant engineering.

Clowe's project management responsibilities include preparing proposals, and technical oversight for plans and reports including subcontracted work, project cost control, and negotiating subcontractor costs. He has managed workloads and tracked contracts to ensure that the services delivered meet client expectations on schedule.

Eventually, Clowe would like to increase his market value by getting a master's degree.

Career Advice

Clowe feels that the skills needed in environmental service are basic science and computer skills, a knowledge of the broad base of environmental and safety regulations, and interpersonal skills. Someone who wants to get into environmental services should get a master's degree in a curriculum that is interesting to them personally, including some engineering courses. While Clowe was able to get where he is without an advanced degree, he thinks that an ideal combination would be a bachelor's degree in civil engineering and an advanced degree in geotechnical engineering, or a degree in chemistry with an advanced degree in geology, hydrogeology, or environmental engineering.

Beyond his BA degree, all of Clowe's training has been through short courses and workshops. He is certified in adult CPR and first aid, data validation, and underground storage system closure. He has taken training in hazardous waste operations, waste minimization, pollution prevention, remediation of contaminated soils and sediments, and Clean Air Act compliance. He obtained all his training on the job; beyond chemistry there was no formal course of study that would have prepared him for what he does.

Clean Air (State Government)
Eric Giroir

Education
BS in biochemistry, Texas A&M University; MS in biochemistry, Louisiana State University; PhD in toxicology, Texas A&M University

Career Path
After receiving his PhD, Eric Giroir was interested in the chemical mechanisms of toxins and worked for a short period as a metabolism chemist in Missouri studying why toxins are toxic at the cellular level. When he was laid off, Giroir looked for work as a chemist, a metabolism chemist, and as a toxicologist. He found a position with the Air Pollution Control Program in the Missouri Department of Natural Resources as a toxicologist, and is continuing in that work today.

Giroir's primary responsibilities include implementing the Clean Air Act Amendments in the state by conducting and reviewing risk assessments; writing a cost-benefit analysis of the Clean Air Act Amendments in the state; developing acceptable ambient levels for air toxins; providing health information on the effects of air toxins; and working with and reviewing Material Safety Data Sheets and toxicological databases. The toxins originate mainly from industrial sources, although St. Louis is categorized as an ozone nonattainment area because of the pollution by automobiles.

Giroir foresees moving up in the hierarchy to a position in the Division of Environmental Quality, working with the Department of Health in explaining the effects of air toxins to the public. He feels that it is valuable to society to have someone reassure the public that they are not endangered by the toxins in the environment.

Career Advice
In addition to technical competence Giroir feels it is necessary to be able to locate needed information in the library and from computer databases to find federal regulations that apply to the state, analyze the information and the data obtained through statistical analysis and good scientific methods, and then communicate to the layperson, the scientist, and the regulator the results of the analysis. Knowledge of computers and computer databases is essential.

Giroir believes that if someone's goal is to work in environmental regulation, he or she should limit the number of laboratory classes and instead take courses such as public health medicine, statistics, political science, and environmental law. Some of these courses are offered in medical schools and schools of public health science. Beyond his degrees in biochemistry and toxicology, Giroir has studied epidemiology, public health medicine, and industrial hygiene.

Clean Water (Corporate Business)
Donald P. Evans

Education
BA in biological sciences, BS in chemistry, BS in geology and geophysics, University of New Orleans

Career Path
Donald P. Evans started his career as a coastal management specialist with the State of Louisiana. He then accepted a position as an environmental specialist at a small chemical company, and went on to become a senior environmental specialist at Gulf Coast Regional Business Unit of Conoco Inc. in Lake Charles, LA. After four years, he assumed his present position as a staff scientist.

Evans ensures that Conoco's operations are in compliance with all federal, state, and local environmental regulations, particularly the Clean Water Act (including such issues as waste water permitting, storm water, and spill incidents). As a registered environmental manager, he estimates that his current position is 70% legal and 30% technical. Recently, he worked with EPA to develop a system for submitting reports to EPA electronically.

Evans sees himself on a career ladder leading toward a management position; he feels he has now reached a level of professional maturity in which he is able to view projects and issues from several perspectives and can make good decisions quickly.

Career Advice
Evans sees the ideal academic path to environmental management through chemistry, chemical engineering, or law. He believes that an environmental scientist should take law courses, particularly environmental law, management, personnel development, and goal-setting. Most important is the ability to work with others since teamwork is essential in today's corporate climate.

Much of Evans' work involves economic analyses and program design/implementation, so computer skills are vital. Since the environmental area is relatively new, he recommends that practitioners should be creative and innovative problem-solvers. Individuals in regulatory work need to be able to handle problems with unanticipated technical and operational variables, and be open-minded about how the variables are interrelated. He stresses the need to be resourceful and thorough, especially in knowing where to find needed information without reinventing the wheel.

Evans feels that chemistry is essential to his work, and that while the geology and biology are not always applicable, they are helpful. He recommends taking engineering and environmental law courses. He learned about the regulations that affect his work from on-the-job training and conferences.

—19—
Science Writing

Science writers fall into two categories: Those who interpret science to the public, primarily journalists, and those who do technical writing and editing for a predominantly science-oriented audience. Both groups share a background in chemistry and require communication skills. They must be able to understand, analyze, and condense scientific information so that the reader can readily understand it.

A journalist must interest readers in what is written. Because journalists write mainly for the public, they must be able to translate scientific terminology so that it can be understood by the public. Journalists sometimes have to subordinate their writing style to the style of a particular publication, and they may also be reviewed by several editors who do not always agree on style and content.

By contrast, the technical writer must be able to satisfy specific needs of an audience, whether writing a safety manual, an environmental impact statement, or an article on using the Internet. The technical writer must understand not only the jargon but the scientific principles behind the subject. The technical writer and editor may be involved in the design and printing of the publication.

Both types of communicators face the same catch-22: You must have writing samples to get a job writing. Most science writers learn by doing and some need further training, either formally or on-the-job, in specialized areas such as government regulations or computing.

This chapter profiles the following examples of science writing: science journalist, technical writer/technical journalist, public communications, public relations specialist, industrial technical writer and editor, and industrial technical editor.

> ### Skills and Characteristics
>
> - A science degree to understand the substance of science and to learn the language
> - Journalism courses or a minor in English can be helpful
> - Short courses in technical editing, media training, public communications for additional training and networking
> - Interpersonal skills to interact effectively with the press and the public
> - Additional experience can be gained through internships, which are generally unadvertised
> - Excellent oral and written communication skills
> - Knowledge of federal and state laws and regulations which govern the environmental field
> - Fluency in foreign languages helps with understanding the English language

Science Journalist

Ivan Amato

Education
BS in chemistry, Rutgers University; MA in history and philosophy of science, Indiana University, Bloomington

Career Path
Ivan Amato became interested in science writing while he was in college and spent a summer working in the mailroom at *Scientific American*. Just after getting his BS degree, Amato returned to *Scientific American* to work in circulation and promotion, and got a good feel for the publishing field. While pursuing his master's degree, he started writing science articles for both his school paper and for a local newspaper. Amato used this experience to get a three-month internship at *Science News*, followed by a stint at the ACS News Service and later as an assistant editor for *Analytical Chemistry*.

Amato then went to *Science News* to cover chemistry and materials science. Three years later, he joined *Science,* where he reported primarily for the Research News section. Two years later he went to the National Institute of Standards & Technology (NIST) as a special projects writer that included ghost writing for the NIST director as well as projects for the World Wide Web. In 1995, at age 33, Amato was one of the youngest recipients of the ACS James T. Grady-James H. Stack Award for Interpreting Chemistry for the Public. Amato has worked on a pilot public television project to promote public understanding of science, one of his many freelance writing jobs. His first book on materials science will soon be published. Amato contributes to a column on the history of the science in the *Materials Science Society Bulletin.*

Amato's writing has included the following topics: chemistry, chemistry's intersection with society, smart materials, light-sensitive chemicals used in cancer therapy, the health impact of garlic, and bond-stretch isomers.

Career Advice

Amato says that in writing for science magazines, the ticket to the business is clippings; editors want to know how well you write, not what degrees you have. He also points out that when you work for a publication you must abide by the editors' quality control rules and write to suit the editor and the editorial policy, not yourself.

Amato feels that the writer needs a basic degree in a science, including working in the laboratory, to understand the substance of the science and especially to learn the language. Also, a course in journalism might help if nothing else than to help one get through the Catch-22 of having to publish before getting published, since editors want to see your clippings before hiring you. His advice is to just start writing and get your foot in the editor's door.

Technical Writer/Technical Journalist

John Borchardt

Education

BS in chemistry with a double minor in English and education, Illinois Institute of Technology; PhD in organic chemistry, University of Rochester (NY)

Career Path

After receiving his BS degree, John Borchardt taught high school for one semester and entered graduate school where he obtained a PhD degree in organic chemistry. After post-doctoral research at Notre Dame, he worked at two companies before becoming a staff research chemist at Shell Development Company.

Today, Borchardt is a chemist at Shell Development Company by day and a technical writer by night. He is the inventor of several commercial products widely used in the paper and oil industries. He holds 29 US and more than 80 foreign patents and has written more than 50 peer-reviewed technical articles and has presented more than 60 papers at technical conferences. He has also written more than 150 articles on various science subjects, computers, career management and job hunting, published in national magazines, encyclopedias, and a major metropolitan newspaper, *The Houston Post*.

Borchardt enjoys writing and sees himself continuing to freelance while working full-time, but he is also thinking about branching out into book writing. When he retires, he plans to write full-time.

Career Advice

Borchardt recommends getting started in writing early, while still working full-time. He thinks there is a market for young writers and suggests writing short articles in the beginning and sending out as many manuscripts as possible. Because it is easier to get published if you have already published, his advice is to sell yourself constantly. He spends as much time writing article proposals and researching articles as he does writing articles. Very few freelance technical writers make a living from writing unless they also write advertising copy, product brochures, and technical manuals. Borchardt does some technical writing for company and product brochures.

Borchardt points out that all chemists have to be technical writers because they write lab reports, progress reports, invention disclosures, articles, and presentations. Technical writing often offers part-time, retirement, or second careers. While he was still employed full-time in his first career, he chose technical writing as a second career to supplement his income and because he enjoys it. He points out that only 30 to 35% of the article proposals he writes result in manuscripts accepted for publication.

Most of Borchardt's material comes from reading and seeing connections that others have not seen, an important ability for an aspiring technical writer. He recommends taking journalism courses or even taking a college minor in journalism to anyone who is interested in technical writing. The Society for Technical Communications (STC) is the largest professional society in the world dedicated to the advancement of the theory and practice of technical communication. STC has more than 20,000 members and 141 chapters worldwide, and offers a clearinghouse for companies who are looking for full-time and part-time technical writers. STC can be reached at 901 N. Stuart Street, Ste. 904, Arlington, VA 22203, or on the Internet at http://stc.org.

Public Communications

Nancy Enright Blount

Education
BA in chemistry, Middlebury College (MA); MS in chemistry, Virginia Commonwealth University

Career Path
Nancy Enright Blount worked at Argonne National Laboratories for six months while in graduate school. After receiving her MS degree, she saw an ad in *The Washington Post* for a science writer at the ACS News Service, and got the job. She began as a staff writer and became a writer for the ACS radio programs, then manager of the News Service, and is now head of the department of media relations/local section public relations in the ACS Office of Public Outreach.

Blount's department includes the ACS News Service and serves as an interface between the ACS and the media, both print and broadcast. They make available to the media information or story ideas that show the relevance of chemistry to daily life, how advances in chemistry improve our lives, and how chemistry helps to solve problems, accurately projecting science to promote the positive contributions chemistry is making to society.

Blount's department also refers experts to the media who can talk about issues that are in the news. They respond to 700 to 800 telephone calls a year from reporters who are looking for information or experts. There is also a program that promotes experts who have interesting things to say on radio or TV talk shows. They train scientists to interact with the media and publish news releases at each ACS meeting, as well as prepare shorter synopses and lists of papers or presentations that might be of interest to the media.

Career Advice
Beyond a chemistry background, communication skills, and people skills, Blount feels that an important skill for public communication is creativity. She defines this as the ability to identify something that has the potential to be of interest to the public, and then to review and translate that scientific material so that the layperson can understand it. The writer must also be able to do feature stories which portray the lighthearted side of science; one example is a story on chocolate cordial cherries and how the cherry and the liquid get inside the chocolate shell.

Blount points out that the public communicator is the mediator between the scientist, who is concerned with detail and precision, and the reporter, who has a minute and thirty seconds of air time for the story, or a press person, who has a deadline and cannot put in all the detail the scientist would like. The public communicator must understand both sides and reach an acceptable balance.

She advises someone who is interested in working in the area of public communications to take a course in science writing. Such courses are now being offered at the University of California-Santa Cruz, Cornell University, the University of Wisconsin, the University of Missouri, and Boston University, among others. Blount also advises getting an internship to learn what public communications people do, and getting clippings that can be used to get a job. She points out that many internships are not advertised; the potential public communicator must be proactive and seek out them.

Public Relations Specialist
Sara J. Risch

Education
PhD in food chemistry, University of Minnesota

Career Path
While studying for her degree, Risch was working in the flavor chemistry laboratory at the university, studying why microwave packaging seemed to impart an off-flavor and odor to the food. Upon getting her degree, she was hired as a consultant by a company that wanted to market microwavable popcorn. Her job was to determine the interaction between the popcorn and the packaging when it was microwaved.

Risch eventually joined the company as director of research and development. Around that time, the FDA realized that their regulations did not address microwave food packaging and decided to hold public hearings on regulating microwave packaging. The company trained Risch in media relations so that she would be able to answer press questions. Over several years she handled the national public relations on the effect of packaging on microwaved foods.

Risch found that she really enjoyed working with the media to get them to understand the science involved. The company gave her opportunities to go before the public and talk about the science of food and packaging. The marketing manager asked her to do a media tour of newspaper, radio, and TV interviews in 25 cities over two years to promote the company's microwavable French fries.

Risch enjoyed getting information out to the public while answering their questions and concerns about microwaving and food. She now runs her own product development company, Science by Design, where she works with a variety of food and packaging companies.

Risch is having so much fun doing what she is doing that she looks forward to continuing her business and helping the public to understand the science and chemistry of food. She has worked on some short courses and will probably do more in the future.

Career Advice

Beyond a thorough understanding of the chemistry and technology of the area you want to promote, Risch says the other requirement to work in public relations is good written and verbal communications skills. She feels she is a scientist who has a gift for explaining science to the layperson. She points out that in public relations, education provides credibility while interpersonal skills help to deal with the media.

Beyond her technical education, the only formal training Risch had was a two-day intensive public media training course she took while working at the food company. The rest she learned on the job. Risch states that a course in public communications would be very helpful to someone contemplating a career in public relations.

Industrial Technical Writer/Editor

Monica C. Perri

Education
BS in chemistry, University of Utah

Career Path

After receiving her BS degree, Monica Perri went to work in a plant tissue culture laboratory for a producer of horticulture plants. She then moved on to become a quality control microbiologist in a microbiology lab for a company making feed additives for the cattle industry. After several years she no longer found laboratory work very satisfying, so she made a career change to become a technical communicator editing and coordinating the production of documents. Perri then secured a position at another company as a communications specialist, focusing on writing and editing.

Perri is currently a senior technical editor with PTI Environmental Services where her work is mostly editing. PTI Environmental Services prepares environmental impact statements, remedial investigations, and feasibility studies for cleanup projects. Perri edits documents written by technically trained people but the primary audience is the public. An important part of her job is to serve as a bridge between the scientist and the public so that the scientific material is understandable to other scientists as well as the nonscientist. She works with the scientists and engineers to maintain the technical integrity of the material.

Perri is very happy with what she is doing and feels that this is the field from which she will retire. She might consider a management position in the future, but for now she is content with her present work.

Career Advice

Perri was able to move into the field of technical writing and editing without any formal education beyond the BS degree. Perri had done volunteer editing of a newsletter and thus had samples of her work to show to potential employers. She also attended short courses and workshops to acquire additional skills and to network with people with similar interests. She is currently pursuing a certificate in technical writing and editing through the University of Washington, a one-year program geared to practitioners.

In addition to the technical background, especially the nomenclature, Perri recommends knowledge of the federal and state laws and regulations governing the environmental field, knowledge that she acquired through study and experience. She must keep up with the literature in the field and had to learn the logistics of document production since part of the job is working with graphic artists and printers. She works with the people involved from the original concept to the final product.

Industrial Technical Editor

Gerald S. Cassell

Education
BA in chemistry, University of Virginia

Career Path
Upon receiving his BA degree, Gerald S. Cassell worked in the Research Laboratories at Eastman Chemical Company (formerly Tennessee Eastman Division of Eastman Kodak Company) where part of his job was writing reports. He was always interested in and adept at writing; in time, he was asked to join the editorial services group to edit technical papers to be published or presented, as well as internal technical reports, according to company guidelines.

He learned editing on the job and was asked to teach a technical writing course at the company. He built the course around grammar and usage rules. This led him to rewrite a technical writing manual for the company which is still in use. He helped develop a registry of new chemicals using *Chemical Abstracts* nomenclature, which he had to learn. He also indexed technical reports for a computer retrieval system to make the information in the reports easier to find.

While Cassell continued his work in technical editing, he slowly moved into the area of information storage and retrieval. As his workload in information services increased, he began editing papers only on request and left the editing of internal reports to authors. In all, he spent 20 years in editorial services and 10 years in information services.

Cassell is now the Research Environmental, TSCA (Toxic Substances Control Act of 1976) and Hazcom (Hazard Communications) Coordinator, as well as the Records Manager. He continues editing and maintaining the Safety Manual and the administrative operating procedures for R&D. He feels that having read much of the material that was written at the company made him more valuable in information services and in his present position.

Cassell thinks he will continue to do what he is doing until he retires. After retirement he could continue technical editing as a consultant, particularly in the environmental area.

Career Advice

Cassell believes that to be a technical editor, it would be helpful to have at the least one course in journalism in addition to chemistry. There are some short courses on writing but few in technical editing. He also advises getting some good books on writing, especially concerning grammar and punctuation.

20

Technical Sales and Marketing

With the increasingly technical content of chemical, pharmaceutical, and medically related products, equipment, and services, there is a greater need for chemists and other technically trained people in sales and marketing. These functions include sales and sales management, market research, market development, and marketing and product management. There are also opportunities for a career in training customers and solving technical problems related to materials, equipment, or services. The lines between all these functions blur in smaller companies where an employee may have to wear more than one functional hat. In larger organizations, career paths can start in one job and move to others within these functions.

A technical salesperson represents technical products to customers who are interested in learning the benefits of these products, including quality and performance measures. These customers often need help in making the right technical and economic choice for their needs, and a technical salesperson with a background like chemistry can provide that help.

Technical salespersons call on customers and potential customers, usually in person, and often follow up by phone, fax, and e-mail. Face-to-face contact is still very important in establishing and maintaining a customer. It is the salesperson's job to know their customer's business and needs, to build a relationship with the key people who make the buying decisions, and to keep them informed about new products and specifications, price changes, quality, and delivery. Salespeople need to be conversant about the technology of the product, its specifications, how the product is used, and how to get help when there are problems with delivery, quality, or perfor-

> ### Skills And Characteristics
>
> - Chemistry or other strong technical background
> - Outstanding interpersonal relations
> - Excellent communication skills, oral and written
> - Organizational skills and the ability to prioritize
> - Business sense to discern the total effort needed to make the sale, the payoff from the sale, and the probability of making the sale happen
> - Problem solving
> - Flexibility
> - Self-confidence when in a remote location and sometimes hostile environment
> - Sales training (usually provided) or experience
> - Leadership qualities
> - The ability to analyze limited information and reach conclusions quickly

mance. Salespeople also provide short- and long-range sales forecasts for their products so that manufacturing can plan production.

Salespeople deal with customers daily, and they are usually the first to hear complaints. Maintaining a positive outlook and demeanor is important. In addition to a chemistry background, technical salespeople need good interpersonal relations, excellent communication, good problem-solving skills, and a high confidence level. Flexibility, initiative, follow-through, and business orientation are also important.

Marketing positions have many titles, roles, and responsibilities. Titles can vary from company to company, especially with the size of the business. Some titles are marketing manager, marketing specialist, product manager, product specialist, market or business development manager or specialist, and marketing or market research analyst, specialist, or manager. In smaller organizations, the title of manager may be used with or without technical marketing professionals reporting to them. Larger organizations may have both managers and specialists reporting to a manager. The marketing specialist in a large chemical company may have more responsibility in business dollar volume than a manager in marketing in a small company. Except where there are staff responsibilities and different levels of decision authority, the elements of the manager and specialist jobs in marketing are similar. Whether for a large or small company, the marketing manager is usually responsible for marketing a product line for his or her organization. Sometimes the sales organization is part of the responsibility, but most times it is a separate group. Since marketing has the overall responsibility for the health of a product line, the position usually reports to a business manager.

The responsibilities can also be very similar for different job titles in different companies. A product manager in one company may do the same job as a marketing manager in another company. In some cases these titles include responsibility for new market or new business development, while in others that is the responsibility of a new business or market development manager.

A product manager or product specialist is responsible for the business and technical health of a product line. This job requires interacting with a number of constituents, including sales, manufacturing, R&D, business, and customers and related management people. They visit customers with salespersons or alone to better understand firsthand the problems and needs of the marketplace. They are responsible for justifying and initiating the development of new products and phasing out low-profit or obsolete products. While the product manager or specialist has a level of authority as part of the job, using persuasion is more effective to accomplish the requested work or changes.

Technical Sales

Thomas G. Tierney

Education
BS in chemistry, St. Mary's College, Winona (MI); MBA, Seton Hall University

Career Path
Thomas G. Tierney's first job was as plant manager at a small natural latex compounding business in Dalton, GA. The business thrived because of changes in the flooring business from woven to tufted carpets, which required a latex backing.

Because of his latex experience, Tierney took a position in 1959 with the Koppers Corporation, which manufactured and sold synthetic latexes. Later,

Koppers bought American Aniline Products, a dyestuff manufacturer. While Tierney was associated with the latex business, he became interested in the dyestuff business. This led to his working in sales of dyestuff intermediates from 1963–69. The dyestuff industry had been highly protected with tariffs since the end of World War I in order to encourage a US-based industry; he found the business and tariff aspects of the industry very interesting. As the tariff protection was eliminated, Tierney recognized the global aspects of this market long before the phrase "global economy" was in vogue.

In 1969 Tierney joined Sandoz selling and buying intermediates for the dyestuff industry. Early on, he set up a computer program detailing the various applications of each intermediate to aid his sales strategy. He also looked for new applications for current intermediates, and was instrumental in bringing Sandoz into the business of using benzotriazole as an oxidation inhibitor for copper sheeting.

Tierney spent the last 20 years of his career buying and selling dyestuff and other chemical intermediates worldwide for Sandoz.

Career Advice
Tierney cites business knowledge, people orientation, strong communication skills, and sales training and experience as important to his sales efforts. He believes that some people are born salespersons and some are made, making a sales career possible for everyone.

He encourages anyone who is interested in selling or working in a worldwide business to acquire fluency in at least one other language. This is especially important now in a global economy. A positive attitude toward foreign countries and cultures is also necessary.

Tierney states that in sales it is also necessary to want to excel, to be able to compete, and to be self-motivated.

Technical Sales Management

Robert DiPasquale

Education
BS and MS in chemistry, Tufts University

Career Path
After Robert DiPasquale received his BS degree, he continued in a graduate program in chemistry but realized that he did not want to work in a lab, so he decided not to pursue a PhD but to embark upon a career.

About this time, his professor was flying back from the Pittsburgh Analytical Conference and was seated next to the vice-president of JEOL USA, a supplier of analytical instruments including mass (MS), nuclear magnetic resonance (NMR), and electron spin resonance (ESR) spectrometers. During the conversation, the professor mentioned that DiPasquale would be a good candidate to sell analytical instruments. This networking contact subsequently led him to join JEOL as an in-house sales coordinator answering phones, handling questions, preparing quotes, and dealing with used equipment, among other responsibilities. He was in a good position to learn the business and was able to complete his MS in chemistry in the evenings.

After a year with JEOL, DiPasquale was assigned a sales territory responsible for NMR, ESR, and mass spectrometer sales. His customers included academic, industrial, and government accounts. His responsibilities included giving one- to two-hour presentations, following up with customers, putting together and presenting quotes, and negotiating with key people and purchasing agents. Sometimes cold calls, which are visits to potential customers without current business or personal relationships, were required to learn about current or future plans or needs for equipment. Entertaining customers or potential customers for lunch or dinner, and sometimes an afternoon of golf or tennis, helped customers and vendors to become more comfortable with each other and develop a relationship.

After a series of expanded territories and roles, DiPasquale was named Assistant National Sales Manager in 1993. He is now National Sales Manager, responsible for six salespersons covering the US, Canada, and Mexico. DiPasquale is responsible for developing, mentoring, assisting, and following up with his sales staff. He often travels with salespersons to visit customers and helps with closing the sale for a piece of equipment. He also participates in the one- to two-day sales presentations for customers at headquarters. For shows or expositions associated with professional meetings like the American Chemical Society or analytical symposiums, such as the Pittsburgh Conference, everyone is available.

Because DiPasquale is new to his current position, he has plenty of challenges and goals to accomplish. In time, he may aspire to the next level of management, but he recognizes that would mean less customer contact, which he truly enjoys.

Career Advice

DiPasquale emphasizes the need for interpersonal and communication skills along with a technical background. The ability to communicate with all levels of management and the ability to quickly grasp an individual's technical level are critical skills.

Because a salesperson usually works out of the home, essential skills include organization, self-discipline, and initiative. In addition, the ability to set the right priorities is a must; sales requires juggling multiple tasks at one time, including the relatively non-productive travel time required to visit customers.

DiPasquale states that when interviewing for a job one should talk with as many people as possible inside and outside the organization to determine whether the working environment is compatible with one's working style.

For DiPasquale, sales offers new challenges every day and the opportunity to meet new people and visit new places. DiPasquale emphasizes the importance of being comfortable working with a variety of people and suggests getting experience working with people prior to a sales career by participating in outside activities such as civic associations, volunteer, or political organizations.

Marketing Research Analyst
Roger Walters

Education
BS in chemistry, Baker University, Baldwin City (KS); international MBA, University of Memphis

Career Path
After receiving his BS in chemistry, Roger Walters joined the Peace Corps. He served in Belize from 1987 to 1989 where he taught college chemistry, established an environmental resource center, organized an international book donation receiving and distribution center, and organized a night school for Spanish-speaking adults to learn English as a second language.

Returning from Belize, Walters joined Buckman Laboratories in 1990 as a technical chemist and then moved into organic synthesis. Buckman is a specialty chemical company making and selling biocides and fungicides worldwide. After taking some graduate classes in management information systems, he pursued an international MBA (IMBA), focusing on international marketing and finance, at the University of Memphis.

After receiving his IMBA in 1995, Walters applied for and was hired for the newly established position of Marketing Analyst at Buckman. His responsibilities include obtaining information on and analyzing the worldwide markets and customers for his company's products; assessing the competitive situation including market shares, strengths, and weaknesses; and establishing and maintaining a market and customer database for the company. Walters has to both search and network to find the information needed. He generates reports to assist management in making business decisions relative to opportunities for selling current products, developing new products, marketing, and staffing. His combined chemistry background, management information systems courses, and business training make an excellent fit for this assignment.

After maturing in his current job, Walters is interested in working in strategic business planning for his company, where plans for new markets or products and investments are developed. Because of his language skills and international experience and education, transferring to a business position outside the US is also a possibility.

Career Advice
Walters cites the need for a chemistry background and chemical business understanding in order to identify trends and their effect on the marketplace. He points out that some changes are driven by outside influences (such as environmental regulations or the economics of low-cost manufacturing processes or locations), while other changes are driven by scientific progress. All such factors need to be understood and analyzed so that appropriate business decisions can be made. Not surprisingly, analytical skills are important for marketing research. Since writing reports and making presentations are a significant part of the job, both written and oral communication skills rank high.

Product Manager

Jeff Oravitz

Education
BS in chemistry and MBA, University of Pittsburgh

Career Path
After graduating from the University of Pittsburgh with his BS in chemistry in 1985, Oravitz joined PPG Industries, located near Pittsburgh, as a Development Chemist in the Coatings and Resins R&D. He moved up as technical group leader in the manufacturing plant in 1987. This job

included some customer responsibility and generated his interest in working with customers.

Oravitz pursued an MBA at the University of Pittsburgh and received his degree in 1992. Subsequently, he was asked to take a position as product research analyst in R&D at PPG. The assignment included market analysis, identifying trends, and forecasting sales, but the focus of the job was measuring or estimating the financial effectiveness of R&D, the earnings value gained or expected from past or future sales of newly developed products, and the business payback of the investment in R&D. Management at PPG moved technical people through this position for increased experience and understanding concerning R&D effectiveness.

Oravitz traded his lab coat for an office with a computer, but not without some concern about mastering the finance skills needed for the job, as well as concerns about moving from the quantitative work in the lab, to the qualitative, less easily defined, forecasts and expectations for the future. Things went well during Oravitz's assignment. In early 1995, he was appointed Senior Product Manager, Industrial Electrocoat group. As product manager, Oravitz is responsible for strategic and tactical planning around his product line, including pricing. He supports new product introductions to the sales group and to the customers in formal presentations. He represents both the technical and business aspects of the current and new products. He devotes some time to looking for new markets for current and new products related to his product line. He estimates he spends about 40% of his time on travel to interact with the salespeople and customers.

Oravitz is relatively new in his position and feels he has much to learn and accomplish before focusing on his next career move. He enjoys the balance of business and technical content in his current position, but recognizes that with career advances, the scale will tip more to the business side.

Career Advice

Oravitz states that a sound chemistry background is a requirement for his position. In his case, the MBA is almost essential for a product research analyst. Making formal presentations, and participating in discussions with customers or sales personnel, require good communication and interpersonal skills. It is also important to be able to size up a sales situation or opportunity with limited information. This requires developing the instincts and skills to ask the right questions, analyze the information or data, and quickly reach a decision about what should be done with the business situation or opportunity.

Based on his experience, Oravitz advises anyone interested in a product manager's position to have broad experience and exposure to different kinds of jobs, and to participate in continuing education, not only for the knowledge but to broaden one's perspective. Internships in industry are a way to get experience in different areas.

New Business Development Manager
W. David Carter

Education
BS in chemistry and BA in religion, University of Richmond (VA)

Career Path
After W. David Carter graduated from the University of Richmond in 1978, he worked on an MS in chemistry. While in graduate school, he worked at a hospital to gain experience for a medical career, and also performed maintenance jobs at an apartment complex. Although he was accepted at medical school, he decided, based on his interim experience and interests, not to pursue medicine.

Carter interviewed at Hercules Incorporated on campus in 1980 and landed an entry-level job as a Technical Service Representative for the Pulp and Paper Chemicals Group in Massachusetts where he set up chemical additive systems at paper mills. This assignment took advantage of his chemistry background as well as his hands-on experience in plumbing, mechanical, and electrical maintenance. Carter soon moved to a satellite office in Maine where he was directly responsible for service and sales. He relocated to a District Sales Office in Wilmington, DE in 1985 where he was an Account Supervisor for customers in four states.

In time he recognized that future promotional opportunities in Hercules were in, or led back to, the home office location, which did not appeal to him. In 1988, Carter went to Stockhausen Inc. as a Technical Sales Representative and moved back to Maine. In 1992 he was asked by a division manager to take a new business development job in the absorbent polymers area. The challenge and opportunity appealed to him and he moved into his current position, Sales & Marketing Manager for the Absorbent Polymer Division. Specifically, he manages the Emerging Technologies Business Center for the Division. The emerging technologies aspect connects him with new business development, although he wears many hats, characteristic of managers in smaller companies.

Carter's current job is to introduce current or modified absorbent polymer products to new markets. He plans, budgets, and executes the plan. Coordinating with specific chemists in the lab who are dedicated to new business development is an important part of the job. He travels up to 40% of his time, making calls on customers or potential customers. While he usually travels alone, sometimes the Customer Service representative or chemist will accompany him.

Career Advice
The first requirement for a business development job is to be fully prepared technically. In his first job as a service representative working in paper manufacturing plants, Carter felt that a chemical engineering background would have been a plus. He does not feel that an MBA degree is necessary, but knowledge about business management is essential, along with continuing education in business. Becoming active in professional organizations related to the business and technical area is also important. Carter is active in Technical Association for the Pulp & Paper Industry (TAPPI) and Paper Industry Management Association (PIMA).

Strong interpersonal relations, communication, and organizational skills are important. Patience is required in new business development; a product may take three or more years to develop and introduce.

Carter cited a number of critical events that led to his current position. His first job gave him an orientation to the paper industry. He enjoyed the hands-on work in the plant that took advantage of his maintenance experience. In working in the plants with operators, maintenance people, and engineers, he developed an appreciation for and the ability to work with a wide range of people.

Marketing Manager
Jim Witcher

Education
BS in chemistry, West Texas A&M (formerly West Texas State University)

Career Path
During Jim Witcher's second year in college he worked part-time for Tech Spray, Inc., in Amarillo, TX in instrumental analysis, progressing to product development in R&D. Upon graduation, he joined the company full-time as a research chemist.

An opening for technical assistance to sales soon became available and Witcher was chosen for the job. In this position he gave sales seminars and participated in two-day training programs for customers. He also served as a new product manager, responsible for coordinating the development and sale of new or modified products to customers. Witcher progressed rapidly and became marketing manager.

Tech Spray's products include solvents, cleaners, and temporary solder masks, which are sold to the electronics industry, specifically printed circuit board manufacturers. Witcher is responsible for all the company's products, pricing, labeling, product literature, and catalogs. Through a New Product Manager reporting to him, he is also responsible for introducing new products. He works with resources inside and outside the company to accomplish this mission. He travels to visit customers and spends 20% of his time visiting customers in the Asia Pacific region, and has direct sales responsibility for Korea, Singapore, Hong Kong, and Manila (Philippines). Direct sales responsibility for a marketing manager in a company is unusual; however, this is another example of the need to wear many hats in a smaller business and in a technical sales role.

While Witcher estimates that he still spends 10% of his time in the lab, he enjoys his current job more than doing exclusively lab work because he feels more in control in his marketing manager position. He sees that his future tends toward business management opportunities.

Career Advice
Interpersonal skills, communication skills, both oral and written, business management abilities, and teamwork skills are all important in Witcher's job. Computer literacy is also very important. A chemistry professor in college took the time to teach Witcher computer programming, which he then did part-time. This sharpened his computer skills, which he has been able to put to good use; Witcher has introduced one-third of the computer programs that his company uses. This effort to reach outside of chemistry differentiated him and established his value to the company in other areas of the business.

He recommends taking business and finance courses, as well as developing complementary personal skills outside of chemistry. Witcher states that exposure to jobs outside the lab is the best route to an alternative career. He sees the value in internships and co-op programs, as well as summer and part-time work, to get that exposure.

—21—

Technical Service

Technical service is part of marketing, but the emphasis is on the technical rather than selling. Technical service personnel usually have access to a laboratory containing equipment that the customer uses to evaluate new products and for customer training. They also have access to the appropriate analytical and physical measurement equipment for staff and customer training. The laboratory is also available to customers to evaluate new products. They also have access to the appropriate analytical and physical measurement equipment needed to solve customer problems or to develop data for product literature or specific customer applications.

Technical service consists of a number of functions. The primary role is to support customers in the use of a company's products, which can include materials, equipment, or services. Technical service people train customers in the use of the products, solve customer problems in using the products, and develop technical information to assist customers in choosing the right product for their needs. Technical service can be the first area to evaluate a new R&D product before taking it to a customer.

In this field, there is a need for strong technical, interpersonal, and communication skills. Business acumen is not as important in technical service as in other marketing and sales jobs because usually the work and priorities are set by the marketing and sales management people. However, handling direct calls from customers or potential customers requires business judgment.

Technical service work offers a combination of technical work and interacting with people. Technical service personnel interface with customers, marketing, sales, and R&D.

The routes to a technical service position vary, but usually start from some technical assignment or experience such as R&D, manufacturing,

> ### Skills and Characteristics
> - Chemistry or other strong technical background
> - Excellent communication skills, both oral and written
> - Problem-solving skills including asking the right questions
> - Good interpersonal skills
> - Organizational skills and the ability to prioritize
> - Flexibility
> - Continued education and learning

quality testing, or sometimes directly after graduation with a chemistry degree. More can be gleaned about the routes and responsibilities from the following profiles of technical service people. While all of them happen to have their PhDs in chemistry, it should be made clear this is *not* a requirement.

Technical Service

Jane J. Janas

Education
BS in chemistry, Marymount College, Tarrytown (NY); MS in chemistry, St. John's University; PhD in organic chemistry, New York University

Career Path
After Jane J. Janas received her BS in Chemistry, she was a supervisor in quality control for The Coca-Cola® Company and then PepsiCo in the New York City area until 1981. During this time, she obtained her MS in Chemistry and decided to pursue a PhD on a full-time basis. After she received her PhD in 1986, Janas went to work for Witco Corporation in lubrication oil additives R&D. She found R&D confining, and began considering other career possibilities.

Janas answered an ad in *The New York Times* for a customer support scientist at Pall Corporation in Port Washington, NY, a supplier of high-quality filtration and separation systems for industry. In spite of having no experience in technical service and filtration, she was hired for the job in 1987 as Senior Staff Scientist.

Janas' responsibilities include visiting industrial customers to evaluate and recommend new or improved approaches to their filtration needs. After a system is purchased and installed, Janas trains the customer and helps start up the filtration system. Through customer visits, she also defines and conveys to R&D the technical requirements for new applications or needs for filtration systems. She supports R&D by reviewing the applicability of media and equipment configurations during development and evaluates them during field tests at customer locations.

Janas likes her job very much. Her career goal is to advance professionally in technical service.

Career Advice

Janas states that the requirement for her job is an advanced degree in a technical subject like chemistry. An informal requirement is to read technical and trade journals related to her market area to keep abreast of market and technical activity and trends.

Not surprisingly, the ability to work with people and communication skills ranked high with her. She emphasized the art of listening and asking incisive questions, stating that it is necessary to depend on and trust your own knowledge and capabilities and be confident. Since Pall's systems are aimed at high value-in-use needs, Janas feels it is important to have a sense of the level of technology needed to solve the problem and the economics of the customer's process and product to see if there is a good match, which calls for some business sense.

Technical Service

Douglas Meinhart

Education
BS in chemistry, University of Rochester; PhD in chemistry, California Institute of Technology

Career Path
During his second year in graduate school, Douglas Meinhart inherited a job maintaining the Nuclear Magnetic Resonance (NMR) spectrometer. He trained other students, kept supplies in stock, and did troubleshooting and repairs.

After a year in a post-doctoral fellowship at the University of Chicago, Meinhart decided not to pursue an academic career. He had found the bench work in graduate school less appealing than his support work with the NMR spectrometer.

While in graduate school, Meinhart had interacted with personnel from JEOL USA Inc., in Peabody, MA who had supplied the NMR, and he expressed his interest in a job with them during a professional meeting and exhibit. This contact led to his joining the company in 1988 as an applications chemist.

Meinhart's title is Manager, NMR Applications Lab, where he spends an estimated 75% of his time supporting customer needs. He demonstrates the equipment for potential customers, trains customers on the equipment they have purchased, and answers technical questions for customers. When a customer has a problem doing an experiment on an NMR, Meinhart may try to duplicate the experiment to resolve the problem. He also prepares customer training materials and instrument manuals. The other 25% of the time is spent working with the engineers who design the instruments to develop new or improved equipment. Here he represents the users and defines their needs based on his experience in working with customers.

His career goal is to advance professionally, to continue building his knowledge base and expertise, and especially to contribute to the development of new equipment. Meinhart likes the flexibility of working in a small organization; there is enough latitude to keep abreast and to contribute to activities in other areas. Meinhart enjoys his work and working with the people in his organization.

Career Advice

Meinhart rates his knowledge of chemistry as the most important requirement in his technical service assignment. He points out that many customers' difficulties with the NMR are related to the chemistry. His knowledge of organic, inorganic, and physical chemistry is used daily.

His work also requires a basic understanding of electronics and radio technology, which is used in analytical instruments. Meinhart acquired this knowledge during undergraduate school, and subsequently he expanded this understanding by working with electronics professionals on the job. Training in computer hardware and software is needed since it is part of the instrument with which he works.

Meinhart lists as key personal attributes patience, teaching capability, good communication skills, problem-solving abilities, and the ability to find information. In addition to good oral communications, good writing skills are needed to write training and instrument manuals. Since a technical service professional works one-on-one in demonstrations and training, good interpersonal relations can make a difference.

Meinhart also states that in his job he can expect to do something different every day, which requires flexibility, and that keeping pace with science is critical to success in this field.

Technical Service

Lisa M. Headley

Education
BS in chemistry, Bucknell University, Lewisburg (PA); PhD in analytical chemistry, University of Akron

Career Path
After receiving a BS in Chemistry in 1966, Lisa M. Headley worked as a chemist at Hughson Chemical Company and then Firestone Tire and Rubber until 1970, when she took a 16-year break to raise her family.

In 1986, when her youngest child entered first grade, Headley decided she wanted to return to the workforce. In preparation, she took some classes at the University of Akron. One of her professors suggested Headley do graduate work with him in analytical chemistry; she liked the idea and completed her PhD in analytical chemistry in 1991. During that time she also was a graduate teaching assistant.

Headley finished her graduate work just as her spouse's company transferred him to Nashville, TN. She found a job in Nashville in a lab doing environmental analyses, and then for a lab doing drug testing. In 1994, Headley answered an advertisement in a Nashville newspaper and landed a position as Senior Staff Chemist with the Boston Weatherhead Division of the Dana Corporation. Her division manufactures and sells sheet rubber and industrial hoses.

One of Headley's responsibilities is maintaining and updating a database on the compatibility and resistance of the 14 elastomers used in sheeting and hoses with over 1700 chemicals and solvents at different temperatures. This information is used by the sales and customer service representatives to answer customers' questions in selecting hose or sheeting. Headley also handles customer questions on compatibility when the answers are outside the database. Headley directs the testing of hoses to generate new data, works with suppliers to get new materials to meet specific customer needs, and conducts some mechanical testing of hoses. She investigates competitive materials, does failure analysis, and has inherited the responsibility for all OSHA and EPA issues, plans, and training at the facility.

As the resident chemist at her location, Headley is also called upon to answer questions and participate in other functions where her chemistry background is needed. She was recently part of a team of people looking at the acquisition of another rubber hose and sheeting manufacturer.

While relatively new in the job, Headley foresees her responsibilities expanding over time, taking on delegated responsibilities. These could be as broad as manufacturing capital spending plans, which her boss, the Director of Technology, now handles.

Career Advice

Knowledge of polymer science and chemistry are required for Headley's position. She is continuing to learn the processes used in manufacturing the company's products by visiting plants and taking a correspondence course in rubber technology. She has taken three short courses on hose applications and technology, including the proper way to put couplings on hoses, a mechanical operation.

Organizational skills, people skills, communication skills, and flexibility are at the top of Headley's list of important skills. She also emphasizes the ability to manage multiple tasks. Headley is called on to give oral presentations to distributors and end-users, and states that good communications skills are important.

Headley advises that industrial experience is worthwhile en route to technical service. She also advises learning as much as possible about materials and materials science, not just chemistry. In retrospect, she would like to have taken some finance and accounting courses so she could have understood the business jargon such as ROA (return on assets) and ROI (return on investment) at the start of her job.

She advises that technical service may not be for the chemist who is happiest in the lab, but points out that she did not know she would like her job outside the lab until she took a chance and tried it.

— 22 —

Technology Transfer

Technology transfer in its simplest form is the process of selling or buying and transferring the rights to a technology from one party to another. The rights are granted in the form of a license agreement. After the license is granted, technological information is transferred which enables the licensed party not only to duplicate, but actually use the technology for commercial purposes. Often the technology is protected by one or more patents or there may be patent applications filed on the technology. Rights are also granted in the license agreement to operate under these current or future patents.

People who work in this area are employed by the buyers or sellers and work on behalf of the buyer or seller to effect a technology transfer. The person representing the buyer defines the technology desired, seeks a source, establishes a value, negotiates an agreement for the rights, assures that the technical information is transferred, and may monitor any obligations under the agreement. The person representing the seller defines the technology owned, establishes a price, defines potential interested parties, markets the technology, finds a buyer, negotiates an agreement, assures that the information is transferred according to the terms of the agreement, and monitors the agreement to ensure that the other party continues to meet the terms of the agreement. Terms may include royalty reports and payments, confidentiality, or rights to any improvements made by the licensee. The technology transfer person is either an employee of the organization or may be an independent consultant hired by the organization.

Technology transfer has been a part of commerce for a long time. Since the mid-1980s there has been considerably more activity in marketing and seeking technology. Developing countries look to technology transfer to

> ### Summary of Skills and Characteristics
>
> - Chemistry or other strong technical background along with scientific breadth
> - A variety of work experience that goes beyond just technical roles
> - Business knowledge and judgment in relation to the technology
> - Market knowledge and marketing skills to target buyers and sellers
> - An aptitude for communication, both oral and written
> - Good interpersonal relations for networking, negotiations, getting information, and marketing
> - Networking skills to help locate technology, sellers, and buyers
> - Problem-solving skills
> - Information gathering skills, the ability to identify and search information sources
> - Analytical ability which includes the ability to make good decisions with a minimum of information and data
> - Good understanding of patents including legal, technical, and business implications and the ability to analyze patents in the areas of the business

make their products and processes more competitive in a global economy. Universities, long a source of new technology, are actively marketing their technology to earn income and research support in a period of declining federal and state funding. Government labs likewise have increased their efforts to develop, market, and transfer technology to the private sector and show that investment in government R&D, beyond that for defense, is worthwhile to the nation's industry, business, and people. Finally, in a global economy, not all the new technology is invented in the US; all of the developed nations are sources of new technology, providing still more opportunities for transfer. These activities have caused an increase of people engaged in technology transfer.

More can be learned from the examples of chemists who are or have been engaged in technology transfer. This chapter includes interviews with people engaged in technology transfer in government, university, and industry settings. There are also opportunities for individuals interested in this field in technology transfer companies or as consultants.

Technology Transfer in the Government
Richard M. Parry, Jr.

Education
BS in agricultural chemistry, University of Rhode Island; MS in biochemistry, University of Connecticut; PhD in biochemistry and food science, University of Nebraska

Career Path
Richard M. Parry's initial interest in chemistry and its relation to agriculture was satisfied when he served as a research scientist for eight years in the US Department of Agriculture Eastern Research Center in Wyndmoor, PA. During this time, he developed an interest in science policy and the public perception of science, but as a bench scientist he rarely interacted with the public. Parry decided to make a career change to science policy but in order to reach that goal, he decided he needed to get some management experience.

Parry applied for and received an overseas job in New Delhi, India, administering research grants in southeast Asia on programs of interest to US agriculture. Because agriculture makes up a greater proportion of the economies in the countries served by the Far Eastern Regional Research Office than in the US, he learned a great deal about production agriculture in this position.

Parry returned to the US as Associate Area Director of the Agriculture Research Service (ARS) in Tifton, GA in 1985 and began to work in all areas of technology transfer, including the development of cooperative research and development agreements (CRADAs) between the ARS and the private sector, patent licensing, and regulatory clearance of products.

Parry moved to his current position as the Assistant Administrator for Technology Transfer in the USDA ARS in Washington, DC in 1994. The ARS is a $700 million in-house R&D operation. His office is responsible for obtaining patents, licensing the technology and patents, and creating agreements with the private sector to develop ARS technologies. He has five Technology Transfer Coordinators who are all scientists responsible for different geographic regions. Parry's office functions as the catalyst to bring together the ARS scientist's expertise and technology with an appropriate private company, which can range from a Fortune 500 to a start-up business. The disciplines involved are broad, including geology, hydrology, engineering, chemistry, genetics, biology, material science, computer software, and human nutrition. The office is concerned with policy in managing intellectual property for public sector research organizations in areas such as ownership of the biology or the chemistry of a gene. Parry works with regulatory agencies because this is an important step in getting some technologies approved and used.

Technology transfer professionals interact with user communities to determine both the user's key problems and the user's desire to resolve a problem. With this information, the odds are greater that a successful research program aimed at one of these problems is going to be useful to someone. It is also easier to attract interest in private sector partnerships with the knowledge that there is a market need. In addition, the market information enables a private company to make business decisions on investing and estimating profitability.

Parry has been in his current position for one year. He feels he has an exciting and challenging job, and his goal is to make his technology transfer program the best one in the federal government. He also wants to use his programs to educate the public on the value of science.

Career Advice
Parry suggests that a PhD is useful for entrée into research and also helps to establish credibility with other scientists. Experience as a bench scientist and in science management, an aptitude for good communications, and an ability to establish relationships with a diverse community are important.

A broad-based knowledge of the sciences is very useful in technology transfer. Parry advises that work experience that combines scientific research in technology with an understanding of the interests of the private sector and user groups for different types of technology is very valuable.

Parry also recommends management experience with staff and budgets before seeking a management job in technology transfer.

Technology Transfer in the University Setting
Robert S. McQuate

Education
BS in chemistry, Lebanon Valley College, Annville, PA; PhD in chemistry, Ohio State University

Career Path

Robert S. McQuate received his PhD in 1973 and was a Post-doctoral Research Fellow at New Mexico State University for a year. He then taught chemistry at Willamette University, Salem, OR. While he enjoyed teaching, he found that the pay was insufficient and decided to change his career path.

McQuate considered a job in the private sector, but because his training and experience was in academe, he felt unprepared to move directly into industry. He decided that a government job with an orientation toward the private sector would give him the necessary experience to facilitate an eventual move into industry. In 1977 he began work with the FDA in Washington, DC as a Consumer Safety Officer concerned with food and color additives, intending to work there five years before seeking a job in industry.

However, a private sector opportunity came sooner than planned. In 1980 McQuate became Senior Regulatory Scientist and Group Leader of Regulatory and Nutrition at the Dial Company in Scottsdale, AZ. In 1983, he learned about a position as Science Director of the National Soft Drink Association in Washington, DC, a highly visible job in a well-funded private sector organization. It was a time when aspartame (the sugar substitute in NutraSweet® and Equal®) was undergoing scientific and regulatory challenges and McQuate's background and experience fit the technical job requirements, and he was hired for the position.

In 1986, McQuate was hired as Executive Director of Advanced Science and Technology Institute (ASTI), Corvallis, OR. He learned about the position through an advertisement in *C&ENews* and was motivated to apply both because his family had enjoyed the quality of life in Oregon, and because the job interested him.

ASTI's goal is to link the research communities of four universities with the private sector and promote technology transfer. The universities include the University of Oregon, Oregon State, Portland State, and Oregon Health Sciences University. Research universities have become more active in marketing and transferring their technology in the last decade. One objective, particularly in the case of state-funded universities, is to transfer technology in order to benefit the state through the employment and taxes generated by either start-up companies or companies that become more competitive through technology. A second objective is to convince the taxpayers of the value of science and university research to them and to their state. A third objective is to develop royalty streams and research contracts for the university to supplement the decreased funding from other sources.

These goals are similar to those for technology transfer associated with government laboratories and research.

McQuate describes his job as marketing and brokering academic research capabilities and technology to the private sector. He looks for opportunities to connect academic researchers to potential customers with two objectives: (1) to attract research dollars and (2) to create licensing opportunities.

To accomplish its objectives, McQuate's office generates general and targeted mailings and newsletters. He has sponsored conferences and workshops related to some of the technologies and research specialties of the universities ASTI represents and attends other conferences where the research strengths of the universities coincide with topics of interest to potential customers. McQuate also visits companies and invites them to visit the university, seeking possible outlets for university technology or for the possibility of research relationships between the two.

McQuate has an agreement in his present job to do some consulting after hours and as long as it does not conflict with his job or his employer's interests. Thus, he has consulted for the past eight years and hopes to concentrate more on consulting in the future.

Career Advice

McQuate feels that his chemistry education has given him a broad background important in his technology transfer job. Written and oral communication skills, including an ability to listen carefully and to understand nonverbal cues, are critical in all of McQuate's activities. He lists the following skills as being important as well: Analytical problem-solving, making decisions based on limited data, and an awareness of business issues and considerations. Continued learning is necessary because of the wide array of technologies with which one deals in technology transfer for a university. Broad experience is an important contributor to developing all of the skills and knowledge needed, and particularly in making good, quick judgments with limited information at hand.

McQuate further cites the importance of networking in developing a career, pointing out that his FDA job and his current position were direct outcomes of his network. McQuate advises chemists generally to expand and hone communication skills and to develop sales skills, since selling research ideas, projects, capabilities, results, people, or yourself are necessary over the course of a career.

Technology Transfer in Industry and Business

S. J. Price

Education
BS in chemistry, Emory University; PhD, Clemson University

Career Path
After receiving his PhD, S. J. "Jay" Price sought a job in the southeastern US since he is from the area, and he became a color chemist at Vanity Fair Mills, Monroeville, AL for two years. When he had been in graduate school at Clemson, Price had contacted a recruiter in Greenville, SC, who had tried to get Price directly into the chemical industry without success. While he was at Vanity Fair, the recruiter found three openings, and in 1973, Price accepted a job in Greenville at Texize, a division of the Morton-Norwich Corporation in consumer products. While at Texize, he progressed through a series of jobs, including Manager of Research Services, Director of New Product Development, and Vice-President, Research and Development.

Dow acquired Texize in 1985 and for a time they ran in parallel with DowBrands, another consumer products business (Ziploc® bags). In 1990 Dow combined and reorganized the two businesses. Price was offered a choice of taking an R&D management job in Midland, MI, or moving into technology transfer at DowBrands. He accepted the technology transfer job and moved to DowBrands in Indianapolis.

Technology transfer has been actively practiced in the private sector for a much longer time than in academe or government. Patent protection has been practiced in industry in the US since the inception of the patent system in the late 1700s. Much of the technology developed by industry has been protected by patents, requiring technology transfer in the form of licensing.

As Vice-President, Licensing and Technology, Price's objective was to move the company more quickly into new products and to reduce new product cycle time. He reviewed and established technology requirements with R&D and the new product development marketing people. He then looked for sources of the technology through databases and networking, and screened and investigated the more desirable ones. He performed "due diligence" studies, which confirms ownership and the right to license the technology, determines that there is no domination by the patents of other parties which could prevent use of the technology or require still another license, and establishes there is no unresolved litigation related to the patents or ownership of the technology. Confirming the workability of the technology is done by R&D or by an independent organization.

After reviews with marketing and business management, and deciding to proceed, Price initiated negotiations for a license or an option to license. He also reviewed some business acquisition candidates where he had some background. Most of the technology licensing was into the business; occasionally as the result of an interest expressed by another party, they licensed to others.

Price has just made a transition into consulting, so his current goal is to learn where his expertise can contribute as a consultant as well as what and where the market is for these strengths.

Career Advice

There are no formal requirements for working in technology transfer. Informal requirements are business, market, and marketing knowledge. Like consulting, a variety of experience prior to working in this area is critical, particularly business experience. Networking, communication, business skills, technical background, analytical skills for problem solving, and patent skills including legal, technical, and business analysis are good personal skills to have. Negotiating skills are needed, and Price recommends taking courses that can teach some of the basic principles.

Price emphasized the importance of business, networking, negotiating, and patent skills. He also recommends taking the Patent Bar Exam early in a career in technology transfer.

APPENDIX

Resources

Books and Articles

"Alternative Careers Lure Chemists Down A Road Less Traveled." 1995. *Chemical & Engineering News* (23 October): 51-55.

Committee on Science, Engineering, and Public Policy. 1995. *Reshaping the Graduate Education of Scientists and Engineers*. Washington, DC: National Academy Press.

"Current Trends in Chemical Technology, Business, and Employment." 1994. Washington, DC: American Chemical Society.

"The Interview Handbook." 1995. Washington, DC: American Chemical Society.

Marasco, Corinne A. 1995. "Is Retraining the Key to Remaining Employable in a Tight Job Market?" *Workforce Report* (April). Washington, DC: American Chemical Society.

Marasco, Corinne A. 1995. "Conducting an Electronic Job Search." *Workforce Report* (December). Washington, DC: American Chemical Society.

Messmer, Max. 1996. "Are You Ready to Job Hunt?" *National Business Employment Weekly* (January 14–January 20): 17–18.

Peterson, C. D. 1993. *Staying in Demand*. New York: McGraw-Hill, Inc.

Peterson, Linda. 1995. "A Step-by-Step Guide to Career Decision Making." *National Business Employment Weekly* (September 10–September 16): 39–41.

Richardson, Douglas B. 1993. "How to Change Careers." *National Business Employment Weekly* (April 23–April 29): 5–6.

Richardson, Douglas B. 1995. "The Perils of Believing Career-Change Fantasies." *National Business Employment Weekly* (August 20–August 26): 21–22.

Richardson, Douglas B. 1995. "Don't Leap Before You Look." *National Business Employment Weekly* (October 29–November 4): 8–10.

Rodmann, Dorothy, Donald D. Bly, Fred Owens, Ann-Claire Anderson. 1994. *Career Transitions for Chemists*. Washington, DC: American Chemical Society.

"Selling Skills, Not Experience." 1996. *Industry Week* (8 January):15–18.

Smith, Joan, and Rose Ann Pastor. 1995. "Interview Tactics for Career Changers." National Business Employment Weekly (May 21–May 27): 21–22.

"Targeting the Job Market." 1995. Washington, DC: American Chemical Society.

"Tips on Résumé Preparation." 1994. Washington, DC: American Chemical Society.

Yate, Martin. 1995. *Knock 'Em Dead: The Ultimate Job Seeker's Handbook*. Holbrook, MA: Adams Publishing.

Electronic Resources

Technology has changed the dynamics of employment through job searching on the Internet, electronic résumés, and scanning technology that affects how companies are selecting candidates to interview and hire. The growing popularity of the Internet has vastly expanded people's ability to make contacts, gather information, and obtain advice that can lead to interviews and, possibly, offers. The appeal of the Internet as a job search tool lies primarily in its low cost; after you pay for access, there are myriad listings and resources available at no additional cost. As the Internet grows and expands, new resources are being added every day.

One way to keep up with the new additions to the Internet is Yahoo!, one of the on-line guides (**http://www.yahoo.com**) to the World Wide Web. Yahoo! catalogues Web pages and links them within hierarchies in addition to offering a search function. Other search engines like Yahoo! are Lycos (**http://www.lycos.com**) and Infoseek (**http://www2.infoseek.com**). Opentext (**http://www.opentext.com**) indexes words and phrases from text in sites, then searches the index to build a list of links to sites that mention your search criteria. Magellan Internet Directory (**http://www.mckinley.com**) lists more than a million Internet sites and rates 30,000 of them while Point (**http://www.pointcom.com**) describes and rates the top 5% of all Web sites.

The following is a collection of career-related sites on the Web. This is only a small selection but these sites contain hyperlinks to other sites of interest. Also, many companies are establishing their own home pages on the Web, so a search of the Web in that regard may be worthwhile.

ACS Job Bank   **http://www.acs.org**

ACS has introduced its own on-line job databank available on the Web. The ACS Job Bank lists jobs as well as provides links to other on-line career assistance programs and corporate home pages. The Job Bank can be found under the Career Services link on the ACS Home Page.

Academe This Week   **http://chronicle.merit.edu/.ads/.links.html**

The Chronicle of Higher Education has long been the best source for academic openings but the print version is difficult to search. The jobs are classified but do not appear in alphabetical order. "Academe This Week" is the on-line version of the current job listings from *The Chronicle of Higher Education*. These are listed on-line every Tuesday afternoon. Job seekers can search using the *Chronicle's* list of job titles or by using any word or words of their choosing. The entire jobs list can be searched by a word or phrase and searchers can limit their search to jobs in particular geographic regions. This site should be searched weekly since the postings are changed every Tuesday.

Academic Position Network   **gopher://wcni.cis.umn.edu:11111/**

The Academic Position Network (APN) is an on-line service accessible worldwide through the Internet. It provides notice of academic position announcements, including faculty, staff, and administrative positions. Included are announcements for post-doctoral positions and graduate fellowships and assistantships. Academic position announcements are transmitted by e-mail or fax and are placed on the network within 24 hours. Institutions pay a fixed one-time fee to post an announcement. APN announcements are unlimited in size and are kept on-line until removal is requested or until the closing date has been reached. APN may be searched by country, state, and institution and may also be searched using a word or combination of words separated by "and" and "or." There is no charge to browse and search the database by authorized Internet users.

America's Job Bank   **http://www.ajb.dni.us/**
The America's Job Bank computerized network links the 1,800 state Employment Service offices. It provides job seekers with the largest pool of active job opportunities available anywhere. The nationwide listings in America's Job Bank (AJB) contain information on approximately 250,000 jobs. In addition to the Internet, America's Job Bank is available on computer systems in public libraries, colleges and universities, high schools, shopping malls, and other places of public access. It is also available at transition offices on military bases worldwide.

Most of the jobs listed are full-time listing and the majority are in the private sector. About 5% of the job listings are in government. The job openings comes from all over the country and represent all types of work, from professional to blue collar, from management to clerical and sales.

There is no charge to either employers who list their job vacancies nor to job seekers who utilize America's Job Bank to locate employment. The services provided by America's Job Bank and each state's Employment Service program are funded through Unemployment Insurance taxes paid by employers.

Current service allows searching for a job through the AJB order file and access to other job banks. In the future, AJB plans to offer alternative methods for searching for a job; job seekers access to information which will help them find work or better jobs; and users other sources of information which will assist them in advancing their careers.

Career Magazine   **http://www.careermag.com/careermag/**
Career Magazine is a comprehensive resource, designed to meet the individual needs of networked job seekers. *Career Magazine* offers:

- **The Job Openings Database**. Every day, job postings from the major Internet newsgroups are downloaded and posted. Postings are searchable by location, job title, and/or skills required.
- **The Résumé Bank**. This gives human resource professionals and recruiters an easy-to-use tool to search and locate résumés of quali-

fied candidates. Candidates can enter résumé information to be placed into the Résumé Bank via an on-line form.

Employer Profiles. Detailed information on employers around the world.

Products and Services to help job seekers manage their careers.

Articles and News to help plan and execute a networked career search.

The Career Forum. This is a moderated discussion area where job seekers can network with others, generate leads, share experiences, and seek advice.

Career Links to other career-related resources on the Web.

CareerMosaic   http://www.careermosaic.com:80/cm/cm1.html

CareerMosaic is an on-line guide to companies and opportunities, made available by Bernard Hodes Advertising, Inc. Using CareerMosaic, individuals can research companies in a variety of businesses, find out what they do, where they do it, and what their environments are like. All the information has been developed by the employer in cooperation with CareerMosaic, so it comes straight from the source. The Jobs section is a link to Usenet Jobs articles in a searchable context. This way, the articles can be searched by topic.

careerWeb   http://www.cweb.com/

careerWeb offers a wide range of resources to assist candidates in their job search process. Through careerWeb, candidates can assess their marketable skills, refine their career search strategy, manage their relocation process, or browse through and apply for available positions.

C.E. Weekly Online   http://www.ceweekly.com/index.html

C.E. Weekly Online is a service of C.E. Publications, publishers of *Contract Employment Weekly*, which publishes job listings for contract technical employment. The site provides three links. Job seekers can search job listings from *C.E. Weekly,* they can get more information about contract employment and résumé writing, and there is an on-line employment office, which contains web pages of contract firms with opportunities.

According to the web site, *C.E. Weekly* is mailed to subscribers every Wednesday via first class mail. Its primary purpose is to furnish subscribers with information about immediate and anticipated contract job openings throughout the US, Canada, and overseas. An estimated 400 contract firms advertise every week. All jobs are for short- and/or long-term contract (temporary) assignments. Subscribers receive a "Directory of Contract Service Firms," which is produced the first week of every year.

C.E. Publications also provides a résumé mailing service. Job seekers can have their résumés mailed out every Friday, monthly, or regionally to advertisers of *C.E. Weekly* or the "Directory of Contract Service Firms." Visit the web site for information on the résumé mailing service and many other services available to candidates looking for contract employment.

E-Span   http://www.espan.com/

E-Span advertises itself "as the first on-line job placement service that does the searching for you!" The system is free to job seekers who fill out a membership form, a user profile, and a résumé sheet. From that point on, candidates simply identify themselves by user name and password. E-Span uses the user profile to match candidates' qualifications against the jobs in the on-line job listing service. When users log in to check listings, they will see only jobs that fit their criteria and interests. E-Span also offers automated e-mailing of jobs matching candidates' qualifications and profiles. E-Span offers complete confidentiality. Since candidates indicate what job listings they find interesting, they are in control over who sees their résumés and who does not. At no time do users of the system search résumés.

Employment Opportunities and Job Resources on the Internet
                             http://www.jobtrak.com/jobguide/

An on-line guide for people who are just starting to incorporate the Internet into their job searches. All of the services listed in this guide are available through the Internet. You may need full Internet access—telnet and ftp in addition to electronic mail[1]—to access most of them. The services listed in this guide can be accessed at no charge; however, some sites may charge a fee to use their search or résumé services.

---

[1]Telnet is one of the basic services which defines the Internet, so if you do not have direct access to things like Gopher and the Web, you can use telnet to connect to public servers for these networks. Telnet creates a connection with a remote machine and enables interaction with it. FTP stands for "File Transfer Protocol" and is the Internet method of copying a file from there to your computer. Anonymous FTP means that you use "anonymous" as the user name when you connect to a site and your e-mail address as the password. (Source: The Riley Guide)

H.E.A.R.T.   http://www.career.com/

H.E.A.R.T. (Human Resources Electronic Advertising and Recruiting Tool) provides the ability to search by position, by employer, or by geographic location. Job seekers can view employment ads and respond directly to recruiters. The service is free to candidates and résumés can only be seen by companies the candidate chooses. Once they have identified positions to apply to, candidates can enter or upload their résumés using a Résumé Builder application.

HiTech Careers   http://www.cyberplex.com/hitech/

This site lists show guides for Career Fairs around North America, as well as information about different high-tech companies and their recruiting needs. The site not only lists upcoming events but also recent events. In case a candidate missed a recent job fair, there is still the opportunity to forward a résumé to the participating companies.

JobCenter   http://www.jobcenter.com/

JobCenter matches candidates with employers. JobCenter maintains a database of résumés from prospective employees and matches them, based on a keyword search, with what the employer is looking for. When there is a match, JobCenter sends a copy of the résumé electronically to the employer. JobCenter then sends a message to the candidate to inform him or her that the résumé was sent for review. JobCenter includes a copy of the job description so the candidate knows exactly what the employer is looking for.

JobCenter also gives candidates the ability to search on-line for résumés and job ads posted to its databases. Candidates get the advantage of automatic notification by JobCenter's e-mail service when a posting arrives that matches what candidates are looking for. Candidates get the added benefit of being able to get on-line and search JobCenter's entire database.

Job seekers can post their résumés on JobCenter for six months for $20. Résumés get maximum exposure through on-line searches, Usenet news feed distribution, and automatic matching sent to candidates and employers via e-mail. If a candidate feels that his or her résumé is not getting the number of replies he or she would like, the match criteria can be altered as many times as the candidate likes. There is no extra charge and only a single copy of the résumé remains on the database. On the other hand, if a candidate is getting too many replies or has accepted a position, the résumé can be deactivated or the candidate can give the remaining subscription time to a friend at no charge.

Job Search and Employment Opportunities: Best Bets from the Net
http://www.lib.umich.edu/chdocs/employment

The authors developed this resource to select from the many employment-related resources and guides found on the Internet and to focus on those sources the authors thought were the best in terms of comprehensiveness, ease of navigation, timeliness, or overall quality. "Best Bets" has links to job postings; places to submit résumés electronically; and career information resources. There is an appendix for those who are new to the Internet with information about common Internet tools, and it offers some tips on navigating the maze of information.

The Monster Board   http://www.monster.com/

The Monster Board advertises itself as the Web's premier career hub, featuring more than 5,000 job opportunities in all fields with over 700 corporations worldwide. Applicants can conduct a search, accessing thousands of career opportunities in all fields. Searches can be conducted by location, industry, company, discipline, or keyword. Candidates can submit an electronic résumé to the Monster Board's national database, make changes to an existing one, and increase their exposure to potential employers recruiting on-line. Employer profiles are also available that contain facts about each company's technology, products, benefits, and work environment. Finally, there is information on which career fairs are coming, job tips, interview techniques, and other career advice.

Online Career Center   http://www.occ.com/occ/

Job hunters can search OCC's employment and resume database by keyword. OCC provides a database, job and résumé files, company information and profiles, and on-line search software to assist both employers and applicants in effectively using Internet. OCC is available via Usenet newsgroups, the Web, Mosaic, Gopher, telnet, and e-mail. OCC is owned, managed, and controlled by employers through a nonprofit association. Employers pay a one-time association membership fee and an annual membership fee thereafter. There are no charges to applicants.

Candidates can browse announcements from one particular company, use keywords to search geographically and by job title, or by posting a résumé through e-mail, where it will remain for 90 days.

Job Listings and Directories

America's Federal Jobs, JIST Works, Inc., 720 North Park Avenue, Indianapolis, IN 46202-3431

A comprehensive guide to more than 300,000 new job openings each year in the federal government.

Chamber of Commerce Directories

Many city and area chambers of commerce publish directories that are similar to the state industrial directories but geographically restricted to areas they serve. These can normally be acquired at nominal cost.

Corporate Jobs Outlook, PO Drawer 100, Boerne, TX 78006, (800) 325-8808

Monthly publication profiling 20 companies per month. A cumulative index is sent with each month's issue.

Directory of American Research & Technology, R.R. Bowker, a division of Reed Publishing, 121 Chanlon Road, New Providence, NJ 07974

This directory includes all known nongovernment facilities currently active in any commercially applicable basic and applied research, including development of products and processes. Most of the entities are owned and operated by corporations, but some university, foundation, and cooperative organizations that do research for industry are also listed. This directory also includes a geographic index, personnel index, and a classification index to R&D activities.

Directory of Directories, Gale Research, Inc., 835 Penobscot Building, Detroit, MI 48226-4094

A guide to more than 10,000 businesses and industrial directories, professional and scientific rosters, directories, databases and other lists, and guides of all kinds. The directory is divided into 15 major classifications with more than 2,100 subject headings including industry, business, education, government, science, and public affairs.

Directory of Graduate Research, American Chemical Society, 1155 16th Street NW, Washington, DC 20036

Lists master's and PhD degree-granting departments of chemistry, chemical engineering, biochemistry, medicinal and pharmaceutical chemistry, clinical chemistry, and polymer science in the United States and Canada. Includes names of faculty members, biographical data, research interests, and titles of their recent publications.

Dun & Bradstreet Million Dollar Directory—Volume I, Dun & Bradstreet, Inc., 99 Church Street, New York, NY 10017

Similar to Standard & Poor's Register, this volume consists of corporations with sales of $1 million or above.

Dun & Bradstreet Reference Book of Corporate Managements, Dun & Bradstreet, Inc., 99 Church Street, New York, NY 10017

Contains data with respect to directors and selected officers of 24,000 companies with annual sales of $10 million or more or 1,000 or more employees.

Encyclopedia of Associations—Volume I, National Organizations of the United States, Gale Research Inc., 835 Penobscot Building, Detroit, MI 48226

A guide to 14,000 national and international organizations of all types, purposes, and interests. Gives names and headquarters addresses, telephone numbers, chief officials, number of members, staffs, and chapters, descriptions of membership, programs, and activities. Includes a list of special committees and departments, publications, and a three-year convention schedule. Cross-indexed.

Job Hunter's Source Book, Gale Research Inc., 835 Penobscot Building, Detroit, MI 48226

Lists profiles of professions and occupations but also information about new companies.

Job Seekers' Guide to Public and Private Companies

Approximately 15,000 companies listed. Human resources and corporate officials' names given. Volumes available for the West, Midwest, Northeast, and South.

National Directory of Nonprofit Organizations, The Taft Group, 12300 Twinbrook Parkway, Suite 450, Rockville, MD 20852

Lists more than 167,000 nonprofits in the United States with reported annual income of more than $100,000.

National Trade and Professional Associations of the United States, Columbia Books, Inc., 1212 New York Avenue NW, Suite 330, Washington, DC 20005

Research Centers Directory, Gale Research, Inc., 835 Penobscot Building, Detroit, MI 48226

A guide to more than 12,000 university-related and other nonprofit research organizations established on a permanent basis and carrying on continuing research programs in agriculture, astronomy and space sciences, behavioral and social sciences, biological sciences and ecology, business and economics, computers and mathematics, education, engineering and technology, government and public affairs, humanities and religion, labor and industrial relations, law, medical sciences, physical and earth sciences, and regional and area studies.

Standard & Poor's Register of Corporations, Directors, and Executives—Volumes I–III, Standard & Poor's, 25 Broadway, New York, NY 10014, (212) 208-8702

A guide to the business community, providing information about public companies of the United States.

Thomas' Register of American Manufacturers—Volumes 1–12, Thomas Publishing Company, 461 Eighth Avenue, New York, NY 10001

Useful in locating many specific product manufacturers, large and small, not listed in Dun & Bradstreet or Standard & Poor's.

ACS Career Services

ACS Career Services exists to enhance the economic and professional status of chemical professionals. To that end, Career Services offers one-on-one career assistance, direct contact with employers, and information on the chemical workforce, trends, and issues affecting employment.

ACS Career Services falls under six categories:

- Career Assistance
- Employment Services
- Workforce Analysis
- Publications
- Workshops and Presentations
- Videos

Services are available to all ACS members—full members, national and student affiliates. For more information, contact:

ACS Career Services
1-800-227-5558
E-Mail: Career@ACS.org
World Wide Web: **http://www.acs.org**